D0090868

This book is a gift of the
Friends of the Orinda Library

Join us at
friendsoftheorindalibrary.org

# HA!

WITHDRAWN

# HA!

The Science of
When We Laugh
and Why

## Scott Weems

BASIC BOOKS

A Member of the Perseus Books Group
NEW YORK

Copyright © 2014 by Scott Weems
Published by Basic Books,
A Member of the Perseus Books Group

All rights reserved. Printed in the United States of America. No part
of this book may be reproduced in any manner whatsoever without
written permission except in the case of brief quotations embodied
in critical articles and reviews. For information, address Basic Books,
250 West 57th Street, 15th Floor, New York, NY 10107-1307.

Books published by Basic Books are available at special discounts for
bulk purchases in the United States by corporations, institutions,
and other organizations. For more information, please contact
the Special Markets Department at the Perseus Books Group, 2300
Chestnut Street, Suite 200, Philadelphia, PA 19103, or call (800)
810-4145, ext. 5000, or e-mail special.markets@perseusbooks.com.

Designed by Linda Mark

Library of Congress Cataloging-in-Publication Data

Weems, Scott.
  Ha! : the science of when we laugh and why / Scott Weems.
    pages cm
  Includes bibliographical references and index.
  ISBN 978-0-465-03170-2 (hardback)—
  ISBN 978-0-465-08080-9 (e-book)  1.  Wit and humor—Therapeutic
use. 2.  Wit and humor in medicine. 3.  Laughter.  I. Title.
R705.W43 2014
152.4'3—dc23
                                                              2013046270

10 9 8 7 6 5 4 3 2 1

*For Katherine Russell Rich, who laughed*

# CONTENTS

# INTRODUCTION

THE JOKE IS DEAD. IT EVEN HAD AN OBITUARY, WRITTEN BY Warren St. John and published in the *New York Times* on May 22, 2005. "The joke died a lonely death," wrote St. John. "There was no next of kin."

The setting was, as armchair poets call it, a dark and stormy night. New York City was being hit by almost twenty inches of snow, with wind gusts exceeding eighty miles an hour and temperatures dropping well below zero. The city was still recovering from an even bigger blizzard just two weeks before, and Mayor Robert Wagner had been forced to declare a state of emergency. Until the weather cleared and bulldozers could push the mess into the East River, New York was closed for business. At the same time, a young comic named Lenny Bruce waited in a hotel on West 47th Street, wondering if anybody would endure the terrible conditions to see his show. Cars weren't even allowed on the roads, so what were the odds they would venture out into the city to see comedy?

It was midnight, February 4, 1961, the beginning of a prolonged death for the traditional joke. By the end of the night, Bruce's career, and indeed comedy itself, would never again be the same.

Bruce had already made a name for himself by performing edgy stand-up routines based on race, religion, and sexual hypocrisy. He didn't tell jokes, and many people didn't find his stories especially funny. Instead, they were shocking, less like comedy and more like social commentary. Bruce wasn't a comedian like Bob Hope or Sid Caesar; his act had little structure and sounded distinctly unrehearsed. Just as jazz musicians hone their craft not by focusing on individual songs but by perfecting use of their instrument, Bruce was becoming master of the riff, the story, and the offhand remark. Carnegie Hall would be his master performance.

The show began with Bruce remarking on the size of the crowd, wondering what would happen if, instead of comedy, he simply performed an extended violin solo. Then his art began, and he ripped into a string of random observations and anecdotes that, if committed to print, would be incomprehensible. He pondered what would happen if Jesus and Moses visited St. Patrick's Cathedral and saw the size of the cardinal's ring. He wondered how, since the earth is constantly rotating, people who die at noon can go to heaven while those passing away in late evening don't go to hell. When feedback erupted over the microphone, he searched the stage looking for the source of the noise, musing how funny it would be if the speakers were simply picking up sound from a kid practicing piano behind the curtain. Like Charlie Parker with a saxophone, or Miles Davis a trumpet, he worked the microphone and improvised about anything that came to mind, drawing huge laughs despite telling almost no traditional "jokes." "There's no right and wrong," he claimed early in the act. "Just my right, and your wrong."

For the next two hours Bruce shared observations about religion, prejudice, and even women with hairy armpits, and though his approach wasn't groundbreaking, it was the first time anybody had performed with such fluidity. Like other comedians of his generation, he rejected the idea of setups and punch lines in favor of a more personal approach, trading one-liners for an angst-filled slurry of words that at times bordered on gibberish. He wasn't the funniest comic of his time. Far from it—much of his humor was lost on the audience for the simple reason that he didn't bother finishing most of his sentences. He wasn't the smartest, either. Rather, he was simply the most creatively idiosyncratic, like the kid in school who could have been voted most likely to succeed if only he cared about the title. He was both a genius and a complete mess.

"All laughter is involuntary," he said during the performance. "Try to fake four laughs in an hour, it'll take you away, man. You can't. They laugh because it's funny. [Switch to stiff, formal voice] *They have had the exposure in the area that he is satirizing.*" In other words, humor happens when we connect with other people and share their struggles and confusions. Indeed, on February 4, 1961, all laughter was involuntary.

Still, the exact moment of death, the joke's final death knell, didn't come until the conclusion of his act. Bruce announced that he wanted to end the performance with a traditional story, one with a regular setup and punch line. People would laugh and jump from the rafters, choral music would celebrate his joy, and his job would be so complete that there would be no need for a curtain call. The joke would be enough.

Nineteen minutes later he still hadn't gotten around to the punch line.

Though the joke eventually elicited huge laughs and applause, the reaction didn't come from the joke itself. That was relatively tame, involving a man sleeping on a plane with his fly open and his privates exposed. No, the audience erupted in huge applause because they recognized that something unusual had just happened. They had witnessed a new form of comedy.

A short while later, Bruce would be arrested for obscenity, and comics like George Carlin and Richard Pryor would take his place as humor pioneers, working up audiences in ways unknown to previous generations. Comedy would remain healthy as ever, though nobody would look at it the same way again.

"I'm not a comedian," Bruce said later. "The world is sick and I'm the doctor. I'm a surgeon with a scalpel for false values. I don't have an act. I just talk. I'm just Lenny Bruce."

〉〉 《《

I'm too young to have ever seen Lenny Bruce perform live, but I love his work and it has often made me wonder: Why do we find things funny? It's a philosophical as well as scientific question: Why do some comments, including jokes, quips, or extended stories, provoke joy and laughter, while others do not? Or, to be more concrete, why do we have the same reaction to a quip made by Lenny Bruce as to one made by Henny Youngman? Youngman was the comic who spoke the immortal line "Take my wife . . . please," the kind of one-liner that's now rare but in its day caused audiences to howl. Humor may have adapted to modern tastes, like other forms of entertainment, but this doesn't explain why something funny to one person isn't to another, or why something that's hilarious in one decade is trite and stale in another.

I believe the answer to these questions lies in the fact that humor is ultimately not about puns or one-liners. Although traditional jokes are now rare thanks to artists like Bruce, humor remains alive and well because it's a process, one that reflects the times and needs of its audiences. It's the social or psychological working through of ideas that are not easily handled by our conscious minds.

As a cognitive neuroscientist with more than a dozen years' experience studying how the brain operates, I have learned that understanding humor requires recognizing the massive complexity of the human brain. If the brain were a government, it wouldn't be a dictatorship, a monarchy, or even a democracy. It would be an anarchy. It's been said that the brain is a lot like the Reagan presidency—characterized by countless interacting modules, all acting independently with only the semblance of a central executive. Political views aside, most scientists would agree with this assessment. The brain is indeed massively complex: parts are connected to other parts, which are then connected to others, but nowhere in the system is there some "final part" deciding what we say or do. Instead, our brains act by letting ideas compete and argue for attention. This approach has its benefits, such as allowing us to reason, solve problems, and even read books. However, it sometimes leads to conflict, for example when we try to hold two or more inconsistent ideas at once. When that happens, our brains know of only one thing to do—laugh.

We often think of the human mind as a computer, one that takes input from its surroundings and acts based on our immediate goals. But this view is flawed. Rather than working in a logical, controlled manner, the brain multitasks. It doesn't break down in the face of ambiguity but, instead, uses confusion to achieve complex thought. When the brain is given conflicting goals or information, it uses that conflict to generate novel solutions, sometimes producing ideas that have never been thought of before. Humor succeeds because we take joy in this process, which is why a bored mind is a humorless mind. We take pleasure in working through the confusion, and we laugh when we've come up with a solution.

One challenge arising from viewing humor as a social and psychological phenomenon is that it's not easily measured. Most scientists prefer to focus on laughter, which is a concrete behavior. As a result, laughter has been relatively well studied; surveys show that we're more likely to be seen sharing laughter than any other emotional response. This means that, on average, we laugh between fifteen and twenty times a day. There's lots of variation, though. Women tend to laugh less as they get older, but not men. And we all tend to laugh more in the afternoon and evening, though this tendency is strongest for the young.

It shouldn't be surprising, then, that our first attempts at understanding humor involved the study of laughter. Aristotle said that humans are the only species who laugh, and that babies don't have souls until they utter their first giggle. As if that wasn't enough, he further claimed that every baby first laughs on his or her fortieth day. Friedrich Nietzsche described laughter as a reaction to existential loneliness. Freud had a more positive view (an unusual role for him), claiming that laughter is a release of tension and psychic energy. The problem with each of these definitions, of course, is that they're useless. There's no way to measure psychic energy or existential loneliness, and there never will be. Perhaps this is why Thomas Hobbes felt comfortable confusing things entirely by calling laughter the "glory arising from some sudden conception of some eminency in ourselves."

Laughter, which we can actually observe and measure, is indeed endlessly interesting, but humor reveals more about our humanity, about how we think and feel, and about how we relate to others. Humor is a state of mind. And that's what this book is about.

⟫ ⟪

*Ha!* is about an idea. The idea is that humor and its most common symptom—laughter—are by-products of possessing brains which rely on conflict. Because they constantly deal with confusion or ambiguity, our minds jump the gun, make mistakes, and generally get muddled in their own complexity. But this isn't bad. On the contrary, it provides us adaptability and a constant reason to laugh.

The reason Lenny Bruce was so funny that night, just like Pryor was a decade later and Louis C.K. is today, is that each found a way to address the prevailing concerns of his time. For Bruce, this involved telling stories about the hypocrisy of sex, prejudice, and drugs, allowing humor to shed light on topics that, in the late 1950s at least, weren't openly discussed. Being funny was how he helped his audience work through living in such volatile times. Indeed, though the traditional joke may be dead (or, more likely, gravely injured), humor remains as healthy as ever because that need for relating with others is timeless.

Over the next two hundred pages I will show that humor is closely associated with nearly every aspect of human cognition. For example, the same processes that give us humor also contribute to insight, creativity, and even psychological health. Studies indicate that the use of humor in everyday settings—for example, when we're responding to e-mails or using descriptive imagery—is strongly related to intelligence. In short, the smarter we are, the more likely we are to share a good joke. We don't even need to be outgoing to appreciate humor. The important thing is being able to enjoy a good laugh.

For years, scientists have known that humor improves our health, and now by viewing it as rigorous exercise of the mind, we understand why. Humor is like exercise for the brain, and just as physical exercise strengthens the body, keeping a funny outlook is the healthiest way to stay cognitively sharp. This also explains why watching Robin Williams's stand-up routines improves our ability to solve word-association puzzles; minds are meant to be constantly worked, stretched, and surprised. Such comedy pushes our brains to make new associations and tackle confusion head-on.

Though in this book we will explore how to incorporate humor more in your life, it's important to note early on that the goal isn't to learn how to make people laugh or tell the perfect joke. This isn't to say that by the end of the book you won't be equipped to be a funnier person. I will show that the key to being funny isn't to learn tricks or memorize jokes but, rather, to gain a firmer grasp of how humor is our natural response to living in a world filled with conflict. Then you will see why comedy follows no simple checklists or rules, and why no single joke pleases everyone. Humor is idiosyncratic because it depends on the one thing that makes each of us unique—how we deal with disagreement in our complex brains.

Some people have argued that there's little point in studying humor because it's too mysterious to understand. The American writer E. B. White even wrote that analyzing humor is like dissecting frogs: few people are interested and the subject always dies in the end. In some ways this is true, since humor is constantly changing and, like a frog on a table, without restraint the subject tends to move on without us. But now scientists are discovering that humor is our natural response to conflict and confusion—a topic absolutely worthy of our attention. What better way to understand what makes us tick than finding out how we cope with uncertainty?

Another common argument against studying humor is that it's as much art as science. Joel Goodman, director of an organization called The Humor Project, once claimed that people learn to become funny the same way a musician gets to Carnegie Hall. That is, they follow the

*Rule of the Five Ps:* they practice—and practice—and practice—and practice—and practice. It's true that humor is so complex, and the causes of laughter so diverse, that no rules apply from one situation to the next. Yet humor has some very clear ingredients, ones that science is just now beginning to reveal. These ingredients explain puns, riddles, and even lawyer jokes. And they all depend on conflict and ambiguity resolution within our highly modular brains.

I will start by introducing you to the latest humor research, showing that it's only through owning indecisive brains that we take pleasure in a cognitively and emotionally demanding world. This raises the *What is?* question of humor: What is it, and why is it so enjoyable? As we will see, humor relies on stages, starting with making premature predictions about the world and ending with resolving the misinterpretations that inevitably result. Without this beginning and end, we don't laugh. Too much in between just muddles the punch line.

The next question is *What for?* What purpose does humor serve, and why do we need such complicated brains? Wouldn't life be easier if our minds were like computers and more predictable? Not at all. First, computers fail all the time, especially when confronted with ambiguity. If a computer gets confused, it must be shut down and rebooted. The brain, by contrast, must keep working even in the face of the unexpected. Second, when was the last time a computer wrote a decent sonnet or composed a catchy song? With simplicity comes a cost.

The last question is *So what?* In other words, how can we use inner conflict to better our lives, and how do we become funnier people? Though this isn't a self-help book, I will show how improving your humor affects your health, helps you get along with strangers, and even makes you smarter. Nearly every aspect of our lives is improved by focusing on humor. This book explains why.

Although my background as a cognitive neuroscientist certainly helped me write this book, I've tried to keep the science accessible to the general reader. One of the most exciting aspects about any emerging science is that at the beginning, everybody is both an expert and an outsider. While many scientists take the subject down some unusual roads—a recent study by researchers at the University of Louisville on the humor of French author Albert Camus comes to mind—the research is still so new it's easy to follow. It also helps that humor has only recently become a legitimate topic of study for academic fields like linguistics, psychology, and sociology. My goal in this book is to act as translator, and perhaps mediator too, pulling out interesting findings from each of these fields. And, by combining their insights, to form a new field altogether—Humorology.

Lastly, I should mention that my goal in writing this book isn't to be funny, though if I occasionally stumble into that too, I don't mind. In fact, I think our overwhelming desire to be funny is the largest impediment to humor research. Humor scientists are notoriously serious about their work, as they should be, because the topic requires precision and academic rigor. But because the subject is humor, many people see the field as an opportunity to tell jokes. And that's a problem. To paraphrase Victor Raskin in his preface to the first issue of *The International Journal of Humor Research,* psychiatrists don't try to sound neurotic or delusional when describing schizophrenia, so why should humor researchers try to be funny? It's a good argument, and one I intend to respect.

Now, on to a laughter epidemic, a disaster movie—and the dirtiest joke in the world.

» PART ONE

# "What Is?"

## THE ELUSIVE CONCEPT OF MIRTH

# 1

## » Cocaine, Chocolate, and Mr. Bean

*There seems to be no lengths to which humorless people will not go to analyze humor. It seems to worry them.*

—Robert Benchley

Let's start with three different instances of laughter— what I call "Kagera," "Stopover at the Empire State Building," and "Titanic." Each is unique, yet together they say something important about what humor is and how laughter is about a lot more than just being funny.

### Kagera

Everybody enjoys a good laugh. But what if you started laughing and couldn't make yourself stop?

Our first laughing event occurred in the Kagera region of Tanzania, then called Tanganyika, nestled along the western shores of Lake Victoria. Located six hours from the nearest airport, Kagera seldom makes the news, which is why it's surprising that the site became host to one of history's most unusual epidemics. Sometime on Tuesday, January 30, 1962, three students at a local missionary boarding school for girls started laughing. Then, as they ran into more classmates, their friends started laughing too, the giggling quickly spilling over to nearby classrooms. Because the students weren't separated by age, with younger and older students sharing rooms, it didn't take long for the laughter to spread throughout the entire campus.

Soon, over half the school's occupants were uncontrollably laughing, almost a hundred people in all. And they couldn't stop themselves, no matter how hard they tried. An outbreak was under way.

Though none of the teaching staff—two Europeans and three Africans—were "infected," the incident quickly overwhelmed the village. Even when adults tried to subdue the laughing girls, the behavior continued. Some students actually became violent. Days passed, then weeks, and when the laughing still hadn't stopped a month and a half later, the school was forced to close. With the students secluded in their homes, the laughter finally settled and the school was able to reopen on May 21, almost four months after the initial outbreak. Then, when 57 of the 159 students became infected with laughter just like before, the doors closed again.

The school wasn't the only location affected, either. Soon after classes were canceled, similar outbreaks broke out in neighboring cities and villages. Apparently, several of the girls, having returned to their nearby homes, brought the laughing sickness with them and infected dozens of others along the way. The epidemic even reached Nshamba, a village of ten thousand people, where it infected hundreds more. No longer confined just to children, the epidemic grew so widespread that the precise number of people affected couldn't even be determined. How could you measure such an event? In total, before the year was over, fourteen schools were shut down and more than a thousand people were overcome by an uncontrollable case of the giggles.

Eventually, the laughter subsided and the epidemic died out on its own, eighteen months after it started. It was as if, for a brief time, the world saw just how contagious laughter can be. The question remains: Why?

## STOPOVER AT THE EMPIRE STATE BUILDING

Our second laughter case study concerns an event that occurred almost fifty years later on the other side of the globe. The location was the New York Friars' Club, just weeks after the attacks of September 11, 2001, and host Jimmy Kimmel was welcoming Gilbert Gottfried to the stage to roast the evening's guest of honor, *Playboy* founder Hugh Hefner. All those before him had avoided any jokes with political or social overtones. Though some referred to the recent tragedy, comments had been short and respectful. Rather than address the prominent topic of the day, they had limited themselves to penis jokes and comments about Hefner's bachelor lifestyle.

Gottfried started his act with a few safe jokes, including one about Hefner needing Viagra. Then he took things a step further, joking that his Muslim name was "Hasn't Been Laid." The crowd laughed, so Gottfried decided to go for broke.

"I have to leave early tonight. I have to fly out to L.A. I couldn't get a direct flight. I have to make a stop at the Empire State Building."

A silence followed. People started to feel uneasy, and several people gasped. Then, the room filled with boos.

"Too soon!" audience members cried. What had been a laughing, supportive audience just moments before was now a room full of judgment.

Gottfried paused. As a professional comedian with over twenty years of experience, he could tell that the crowd had turned on him. He had crossed a line. Some performers would have acknowledged the mistake and returned to their safe material. Others might have simply left the stage. Gottfried went in a different direction.

"Okay, a talent agent is sitting in his office. A family walks in: a man, woman, two kids, their little dog. And so the talent agent asks, 'What kind of act do you do?'"

I wish I could tell you the rest of the joke. I really do. But there's no way you'll ever see it in print, just as you won't see it in recordings of the roast. The joke's depravity either broke all the cameras or scared Comedy Central so much it burned the tape shortly thereafter. The joke, named for its punch line *The Aristocrats,* involves scatology, violence, and even incest, and though it has been around for decades, it's almost never told in public because it's literally the dirtiest joke in the world. After the setup—that a family walks into a talent agent's office to describe a proposed act—the joke goes on to describe the most obscene performance possible, filled with sex and unspeakable taboo. The punch line, that the performers give their act the very proper title *The Aristocrats*, is less a traditional punch line than an opportunity to share a revolting setup.

Though the audience was wary at first, as the obscenity escalated Gottfried's commitment eventually won them over. Soon the crowd was roaring and many attendees, themselves performers with exceedingly high comedic standards, fell to the floor laughing. By the time the joke was over, some were guffawing so loudly that, as one journalist put it, it sounded as if Gottfried had performed a collective tracheotomy on the audience. The performance was so memorable that someone made a movie about the joke, with Gottfried's performance as the climax, titled *The Aristocrats*. I implore you to look it up if you aren't easily offended.

Gottfried killed the rest of his performance and, at least partly because of that joke, is now considered a New York legend. He's a comedian's comedian, and the joke helped a lot more than just his career, if you take his word for it.

"The only reason America is standing today," he claimed in an interview many years later, "is because I told that joke at the Hugh Hefner roast."

## Titanic

Our final laughter case study is more personal. Around Christmas of 1997, my wife Laura and I went to see the movie *Titanic* with my parents. It was a stressful time because we had just moved to Boston to start new jobs. But we wanted to see our families for the holidays, so we packed up the pets and drove to Florida to visit my parents, and as often happens when visiting family, by the second day we were already running out of things to do. Agreeing on a movie was difficult, but in the end we didn't have much choice. One film was dominating theaters, with new showings starting nearly every hour. We were about to watch a movie about an iceberg.

I don't mean to give away any spoilers, but there's a scene near the end of *Titanic* where Leonardo DiCaprio is freezing to death next to the sinking ship while Kate Winslet clings to a floating piece of debris. Leo is about to die, Kate has a renewed interest in life, and Kathy Bates is complaining in the distance that *somebody needs to do something!* Over two hours of love story has built up to this moment, and director James Cameron is playing it for all it's worth. As I watched the tragic scene, I turned to look behind me and saw that every person in the audience was crying. Women and men alike were sobbing into their shirt sleeves, including my father who to this day claims he simply ate too many Red Hots.

Then I looked to Laura. She was laughing.

Now, I don't want to make my wife sound insensitive. She cries a lot, or at least a normal amount for a woman her age. She can't even listen to music by Sarah McLachlan because it reminds her of the SPCA's anti-animal-cruelty commercials. But there was something about the ridiculousness of the situation that made her lose control at the movie theater that evening. She tried to hold in her emotions, but the more she struggled to conceal them, the more they burst through. People around us started to become irked, which only made things worse. I asked Laura what was going on.

She waited several seconds before answering.

"Yo, Adrian," she whispered in my ear, a reference to the line from the *Rocky* movies. Apparently, the scene in front of us had made her think of that love story from Philadelphia, while everyone else was still grieving in the Northern Atlantic. She even slurred her syllables, just like Stallone. Except in the current movie, the characters were slurring because they were freezing to death. In *Rocky*, that was just how they spoke.

My parents were not amused.

» «

My sharing these three incidents may seem a strange way to start a book on humor. After all, only the second one involves a traditional joke (however loosely defined), and as noted, it's so obscene I can't even repeat it here. In Laura's *Titanic* example, only a single person laughed, and everyone nearby thought the behavior both inappropriate and disturbing.

My hope is that, by the end of this book, you will look at humor differently—no longer in terms of jokes but instead as a psychological coping mechanism. This is exactly what the three case studies have in common. In the pages that follow, we'll see why humor, though it takes many different forms, can't be reduced to a single rule or formula. Instead, we must see it as a process of conflict resolution. Sometimes that conflict is internal, as with Laura's breakdown at the movies, and sometimes it's social, as with Gottfried's joke. At other times, as with the children of Kagera, it's a combination of both—the only way to deal with a life in turmoil.

## WHAT IS HUMOR?

For many of us humor is synonymous with being funny. Someone who cracks a joke or makes us laugh is considered humorous, and having a sense of humor means being quick to recognize a punch line or share an amusing anecdote. Yet, closer scrutiny shows that humor isn't always so straightforward. For instance, why are some jokes hilarious to some but grossly offensive to others? Why do villains laugh as they're conquering the world, or children laugh when they're tickled? Why is it (to paraphrase Mel Brooks) that when I fall down a manhole it's funny, but when you do the same thing it's tragic?

To pick a specific example, consider one of the greatest comedies of all time, as judged by the American Film Institute: *Dr. Strangelove or: How I Learned to Stop Worrying and Love the Bomb.* In that movie soldiers are shot, men commit suicide, and the entire world is ultimately destroyed by nuclear war. Yet, it's considered humorous because all the death and destruction are intended to be ironic. The film almost completely lacks traditional jokes, yet the pointlessness and futility it portrays make us laugh because we have no other way to respond.

Under the right circumstances almost anything can make us laugh, which is why humor should be considered a process, not an outlook or a behavior. It results from a battle in our brains between feelings and thoughts—a battle that can be understood only by recognizing what brought the conflict on. To understand why, let's revisit the three scenarios that opened this chapter.

In the first scenario, the outbreak at Kagera, we see an important distinction regarding humor: not everything that makes us laugh is funny. The children who were affected, though laughing on the outside, reported extreme stress—and desperately wanted to stop. One interpretation is that they experienced mass hysteria brought on by the stress of massive social change. The prior December had marked the country's independence from Britain, and the school itself had just abandoned racial segregation, integrating its students at a time of intense cultural sensitivity. Add to this the fact that the students were adolescents, many just entering puberty, and the pressures were immense.

But this doesn't explain: Why *laughter?* History is filled with social and cultural change, yet epidemics like this are rare, and when they do occur the behavior is usually complex. In sixteenth-century Europe, for example, groups of nuns once spontaneously erupted in convulsions while mimicking the sounds of local animals. Dozens of convents were affected. In one, the occupants uncontrollably meowed like kittens; in another, they barked like dogs. In a convent in Xante, Spain, they bleated like sheep. Scientists agree that stress brought on these outbreaks too—specifically, stress caused by strict religious indoctrination and widespread talk of witchcraft. The nuns had felt so threatened by spiritual possession that they began adopting the very behavior they had been warned against.

It would be easy to say that the Kagera children simply experienced a breakdown. Asked to live in two worlds at once—not British or African, not black or white, not even adult or child, but a combination of each—they failed to cope. But laughter isn't a breakdown. Convulsing on the floor while meowing like a kitten is a breakdown, but laughter is something entirely different. It's a coping mechanism, a way of dealing with conflict. Sometimes that conflict comes in the form of a joke. Sometimes it's more complicated than that.

Consider the story of Conchesta, one of the children affected by the outbreak. As a teen she, too, had been overcome with laughter during the epidemic, and when asked about the laughter later she claimed that it mostly struck girls who "were not free." When the reporter asked if Conchesta felt free, her answer was immediate.

"When you live with your parents and you're that age, no one is really free."

Conchesta's story reveals a brain mired in conflict. At the time of the outbreak she had been seeing a nearby boy, but like most pubescent girls she was prohibited from spending time alone with members of the opposite sex. Normally an established courtship process would have allowed the relationship to bloom under close scrutiny, but Western values had changed everything. Catholic and Protestant churches began offering villagers money for joining their congregations, bringing with them new rules for sex and marriage. Clans disintegrated, and so did established structures for young, pubescent girls to find possible mates. Conchesta wasn't free at that time because she didn't know who she was anymore. Her brain was in a state of transition.

Conchesta's story was typical among the Kagera children, but her explanation for the outbreak was less scientific. Before the outbreak, she said, the village had been struck by an infestation of caterpillars, which grew mostly in nearby fields. These caterpillars, though individually harmless, had a history of arriving in swarms in late winter and early spring. They could destroy an entire crop in a matter of days, so their appearance was anything but welcome. Children were warned to stay clear of the fields for fear of disturbing the visitors and drawing their ire. Those struck by the laughter, according to the legend, had ignored the instructions and crossed a field, killing several of the caterpillars and angering their spirits. The laughter was those spirits' retribution.

Nobody thought to ask whether Conchesta was one of the children who had illicitly crossed the fields, or to associate the outbreak with another unique aspect of the caterpillar—that it, too, inhabits two worlds at once. At birth it's a larva feeding on leaves and grass, a destructive force capable of wiping out entire crops in just a few days. But inside its cocoon, it's an African Armyworm moth, waiting to emerge and fly to distant lands hundreds of kilometers away.

〉〉 〈〈

In the second scenario, Gilbert Gottfried told the most obscene joke in the world to an audience already wary of offensive material, yet he succeeded because his joke communicated a sensitive and subtle idea— one that endeared him to the audience. The idea here is that obscene jokes are intended not to offend but, rather, to question what it means to be offended in the first place. Obscene humor challenges accepted norms and makes us laugh not *despite* its depravity but *because* of it.

Humor—especially offensive humor—is idiosyncratic. People have different thresholds for what they find offensive, and they vary widely in their responses when that threshold is crossed. Still, Gottfried's brazenness in tackling prevailing sensitivities head-on was impressive. Had he simply told his audience to chill out, he would have been booed off the stage. Had he spewed vile and filth outside the context of a joke, the audience's reaction would have been even worse. Humor provided him a tool. And he used it expertly.

Gottfried's joke also reveals the jointly psychological and social nature of humor. There's an old saying that if you want to make a point, tell a story—but raising several points at once requires humor. Cutting-edge humor never involves just a single message. There's what the humorist is saying, and all the rest left unspoken. When Gottfried told *The Aristocrats* joke, he wasn't celebrating perverseness. Rather, he was sharing his desire to be funny while also remaining respectful to the recent victims of 9/11, and the only way to do both was to have his audience struggle with the same challenge. That required showing them that even the vilest words don't physically hurt anyone.

Even animals use humor as a tool for diffusing tense situations. For example, chimpanzees bare their teeth in laughter during friendly interactions, especially when meeting strangers and forming new social bonds, and dogs, penguins, and even rats have all been shown to give hearty chuckles during rough-and-tumble play. Consider, for instance, a study conducted by members of the Spokane County Regional Animal Protective Service. They recorded the grunting noises made by shelter dogs during play, noises that seemed eerily like laughter. When those same noises were broadcast over speakers in the shelter, the dogs not only became more relaxed but also played more. They wagged their tails and generally acted as though they were relaxing in a comedy club rather than being confined in a kennel.

Our similarity to other species isn't limited to laughing, either—some animals even demonstrate a rather provocative sense of humor. A case in point is the chimp named Washoe, one of the first animals to learn American Sign Language. Washoe was raised by primate researcher and adopted parent Roger Fouts, and according to one frequently repeated account, one day Washoe was sitting on Fouts's shoulders when suddenly and without warning he began to pee. Of course Fouts was disturbed by the incident, as anyone would be under such circumstances, but then he looked up and saw that Washoe was trying to tell him something. He was making the sign for "funny." The joke, apparently, was on Fouts.

〉〉 《《

The third scenario asked why Laura would crack a joke while watching the closing scenes of *Titanic*. We could ask Laura herself, but psychology suggests that doing so would provide an unreliable answer. Laura probably doesn't know any more than we do. We can only look at her actions, which brings us to "Yo, Adrian!"

As we've seen, humor is often thought of as involving jokes, even obscene ones like Gottfried's. This, however, was a different situation entirely. Laura laughed while surrounded by dozens of crying people, none of whom thought her actions were appropriate for that moment. In fact, several people shushed her, including her mother-in-law, something that would never have happened had we been watching a comedy. There was no social expectation of laughter, and no punch line, either—only an embarrassed wife and a crowd of angry moviegoers.

The American Film Institute lists "Yo, Adrian" among the most influential lines in movie history, though it isn't recognized for being deep or meaningful. On the contrary, it's just one of those phrases that comes out of our mouths. When the *Rocky* movies were first released, everybody was mimicking Stallone's slurred "Yo, Adrian." The line is even repeated in the sequels, and in each case it's portrayed as an honest, unsentimental call to Rocky's love. This isn't to say it's a simple or meaningless line. Far from it—it's genius. After Rocky survives his fight with Apollo Creed, his call to Adrian is a touching climax. Punctuating the scene with a short, slangy line is real life. It's the noticeable absence of sentimentality.

I can't say what Laura felt, but obviously she wasn't moved by the DiCaprio character's demise. My guess is that her brain needed a way to resolve the conflict between watching a tragic death on-screen and feeling like her emotions were being manipulated with a sledgehammer. "I just saw all the people crying and for some reason I imagined Sylvester Stallone, I mean Rocky, out there in the water too, yearning for Adrian," Laura told me afterward. "And I asked, *What would Rocky say?* There was no getting it out of my head at that point. I wanted to cry, I really did. I just really wanted Rocky to be out there too."

In Laura's reaction we see another important psychological principle governing humor, which is that we react to humorous situations everywhere, and we've all laughed at situations that only we thought amusing. Laura was the only person laughing in the theater because only she found the overwhelming sentimentality entertaining, her brain struggling to resolve her opposing emotions about what was happening on-screen. On the one hand, she experienced sadness while watching hundreds of people tragically drown, including the male lead character. On the other, she could see director James Cameron treating the emotional climax in front of her the same way he treated the action-based climaxes of his earlier films, *Aliens* and *The Terminator*—with nonstop fury. That's a lot to ask of anyone.

It may seem that each of our three laughter case studies has moved further and further away from the traditional concept of humor. They have, but as we've seen, humor isn't just about being funny; it's also about how we deal with complex and contradictory messages. It helps us resolve confusing feelings, and even connect with others in times of stress. Laughter is simply what happens as we work through the details.

## THE ELUSIVE CONCEPT OF MIRTH

Imagine that it's the middle of the twentieth century and you have just volunteered to participate in a study on humor. The researcher wants you to view a series of hand-drawn cartoons. Act naturally, he says, and laugh only when the feeling strikes you.

The first cartoon depicts a man casually raking leaves, next to a buxom woman tied to a tree. There's no explanation, just a woman who looks irate and a man who appears happy to be experiencing the outdoors without his partner able to interfere. The second cartoon shows a man and a gorilla walking into a pet store next to a sign reading "Pets bought and sold." In the second frame, the gorilla walks out of the store holding a stack of money in his hands. The third cartoon is from *The New Yorker* and depicts two skiers, one facing uphill and the other down. Behind the downhill skier is a set of tracks passing around a tree. Except that the path of the left ski passes to the left of the tree and the other to the right. The uphill skier looks on in befuddlement.

None of these cartoons is particularly funny, but you chuckle at the second one—the one with the gorilla—as well as at the last one with the skiers. You notice that the researcher is taking copious notes, and when the test is complete you ask him how you did. He says you show signs of anxiety. Why? He replies that the first cartoon, the one featuring the tied-up woman, is a "sensitive stimulus." Anxious people and schizophrenics tend to be disturbed by the thought of involuntary restraint and thus don't laugh at that one, whereas normal people find it amusing because they recognize that the violation is minor and that the man is just using an unusual, and potentially humorous, means of enjoying a sunny day. The researcher goes on to tell you that the other two cartoons, the ones with the gorilla and skier, aren't particularly provocative, so it's interesting that you found those amusing. Normal people typically require that their humor make them a little uncomfortable, and these cartoons shouldn't satisfy that need.

But don't worry, he adds. It's only one assessment.

You have just taken the Mirth Response Test, a humor tool from the mid-twentieth century that was once popular enough to be featured in *Life* magazine. It's based on Freud's theory that humor is our way of resolving inner conflict and anxiety. According to Freud, we constantly desire things such as food and sex. At the same time, our anxieties keep us from acting on these desires, leading to inner conflict. Humor, by treating these forbidden impulses lightly, allows us to relieve inner tension—in other words, it permits us to express ourselves in previously forbidden ways. This is why successful jokes must be at least a little provocative. Too much anxiety and we withhold the laugh. Too little and we don't laugh because our humor system isn't engaged at all.

The funniest things are those right in the middle. Individuals suffering from schizophrenia or high levels of anxiety generally enjoy only the milder cartoons because they have enough stress in their lives already. Everybody else prefers more of a middle ground.

Though few scientists take Freud seriously now, most recognize that there's at least a kernel of truth in his theory. Jokes that fail to make us at least a little uneasy don't succeed. It's the conflict of wanting to laugh, while not being sure we should, that makes jokes satisfying.

We laugh at what forces us to integrate incompatible goals or ideas that lead to confusion, doubt, and embarrassment, but the form of what brings on these reactions varies widely. For example, there are riddles, puns, satire, wit, irony, slapstick, and dark humor, to name just a few. Asa Berger, prominent humor researcher and author of more than sixty books on such topics as the comic book industry and Bali tourism, identified as many as forty-four separate types of humor. Realizing that this number was getting unmanageable, he went on to group them into four categories: linguistic, logical, active, and identity-based. Slapstick, for example, is an active form of humor. Caricature focuses on identity.

Future chapters will explore some of these humor types in greater detail, but for now, let's focus on slapstick as an example. Slapstick humor involves exaggerated violence, often in the context of crashes and collisions that occur outside the boundaries of common sense. In other circumstances such violence would be frightening, but with slapstick it's humorous. Why? Because when the Three Stooges strike each other with bats, they do so with exaggerated motions and the understanding that the violence isn't intended to injure or maim. It's still violence, but it's harmless, a perplexing paradox leading to laughter. If the violence were realistic, it wouldn't be funny, which is why striking a stranger with your car is a felony. Doing the same thing to Johnny Knoxville wearing a chicken suit will get you on television.

Even with all this variation, humor's effects on the mind are the same for everybody—chemicals flood the brain, resulting in joy, laughter, or both. Though many people think of the brain as an electrical machine, this is a misconception. Individual neurons internally rely on electric polarization, but the connections between neurons are almost always chemical. This is why certain drugs can have strong effects on our thinking—they're made up of the same substances as those used by the brain to convey messages.

Dopamine, the neurotransmitter most closely linked with humor, is often considered the brain's "reward chemical." That's why it has also been linked with motivated learning, memory, and even attention. Food and sex stimulate the brain to increase available dopamine too, whereas dopamine deficiencies lead to impaired motivation. Cocaine also increases dopamine availability in the brain, which is why it's so addictive; after the initial high, the user is left desperately wanting more. Chocolate does largely the same thing, just not as strongly.

We know that dopamine is important for humor because we're able to look at people's brains as they view jokes and see what happens. This is what the neuroscientist Dean Mobbs did at the Stanford Psychiatric Neuroimaging Laboratory. Specifically, he showed subjects cartoons while they were being monitored by a magnetic resonance imaging scanner, known popularly as an MRI. Half of the eighty-four cartoons were chosen for being particularly funny, while the other half had the funny parts removed (see Figure 1.1). His goal was to see what parts of the brain became active during the funny trials but not the others.

Mobbs saw that subjects' brains became highly activated for all the cartoons, but one subset of structures responded solely for the funny ones—namely, the ventral tegmental area, the nucleus accumbens, and the amygdala. What do those brain regions have in common? They're key components of what scientists call the dopamine reward circuit, which is responsible for distributing dopamine throughout the brain. In response to unfunny jokes, we not only fail to laugh, we miss out on the joy. That joy comes in the form of dopamine.

The dopamine reward circuit is one of the brain's most misunderstood regions, partly because it does so much. It's important for emotions as well as memory, and has been linked with classical conditioning, aggression, and even social anxiety. It's so important because

FIGURE 1.1. One of the cartoons shown to subjects while they were monitored by an MRI. For the "funny" version, the unaltered cartoon was used. For the "unfunny" version, the alien was removed and the man made the remark about hallucinating to himself. Only the funny version led to activation in dopamine centers of the brain. Copyright BIZARRO © 2013 Dan Piraro, Distributed by King Features.

reward is how the brain keeps itself going. We often think of rewards as things we are given, rather than give ourselves, but the brain doesn't work that way. To keep us making good decisions, it gives rewards to *itself* all the time. That's why emotion is such a key element in successful decision making. Dopamine is the currency that allows the brain's government to operate.

It's worth taking a moment to recognize this important fact—humor taps directly into the brain's pleasure-production system. To explore this concept, let's compare two studies, each examining very different phenomena. The first was conducted at McGill University in Montreal, Canada, where ten musicians listened to pieces of music identified as being so emotionally moving as to induce chills—that shivers-down-the-spine feeling that accompanies intense euphoria. For each musician just such a piece was chosen before the experiment began, and then researchers identified the brain regions responsible for the feeling while the musicians listened to their songs. The culprits? Not surprisingly, these were the amygdala and the ventral striatum of the dopamine reward circuit, as well as the primary region they are connected to: the ventral medial prefrontal cortex.

Subjects' brains were monitored in the second study too, but this time the experimenters showed video clips of the British television show *Mr. Bean,* starring Rowan Atkinson. This series, which focuses on the physical comedy of Atkinson as he solves everyday problems with child-like confusion, is unique in that it features almost no dialogue. This allowed subjects to be shown matched funny and unfunny bits whose only difference was their inherent level of humor. Half of the videos were taken from the show's funniest bits, while the other half included no humorous elements at all, and subjects were instructed to mimic laughter even when they didn't find them funny.

The brain area most active during the funny parts, but not the others, was the ventral medial cortex, the primary target of the dopamine reward circuit. This is the region responsible for differentiating true laughter from pretend, the same one that apparently gives some of us chills when we listen to Samuel Barber's *Adagio for Strings*.

From these findings you might suspect that dopamine is one of the most important chemicals in the brain, and you'd be right. Scientists have even proposed something called the Dopamine Mind Hypothesis, which states that increased reliance on dopamine helps explain our evolutionary separation from lower ape ancestors. According to this theory, when *Homo habilis* took up meat eating around 2 million years ago, brain chemistry began to alter. Dopamine production skyrocketed, and so did the incidence of cognitive and social processes depending on this chemical, such as risk taking and goal-driven behavior. In short, dopamine made us who we are—physical and intellectual thrill-seekers, always on the lookout for some new way to improve our lives or make ourselves laugh.

We have proof that dopamine is key for animal humor, too, most notably from Northwestern University's Jeffrey Burgdorf. Not only did he learn how to tickle rats, he was able to set up recording devices to hear their laughter. Apparently, one tickles a rat by scratching its belly, causing it to emit high-pitched screeches at around 50 kHz, well outside human hearing range but easily audible to a rat. Burgdorf showed that rats respond to tickling the same way as humans, running in anticipation to tickling fingers and sometimes laughing even before any contact is made. Stroking (i.e., petting) the rats doesn't elicit the same reactions, and neither does holding them. Burgdorf further demonstrated that older rats respond less to tickling than young ones, as with humans, and that young rats who are lonely as a result of being isolated from peers are the most prolific laughers of all.

But more importantly, Burgdorf showed that tickling wasn't the only thing that brought on laughter in his rats. Inserting electrodes in their dopamine-producing centers achieved the same result. He even trained rats to stimulate their own brains by pressing a bar, delivering a current to their dopamine centers and causing them to laugh even without any tickling. Administering dopamine-promoting chemicals directly into the rats' brains had similar effects.

Apparently rats aren't so different from humans, which suggests that laughter might have been around for a very long time. Perhaps it developed to help women like my wife cope with excessive sentimentality, and girls like Conchesta deal with political and social upheaval. For Gilbert Gottfried, it may even have helped prevent a sensitive audience from booing him offstage. Now that we're no longer able to resolve confusion by picking fleas from each other's fur or beating each other with sticks, our humor has evolved just as we have. And that evolution has taken some very broad turns.

## THE FUNNIEST JOKE IN THE WORLD

Legend has it that there are only five jokes in the world. I suspect this myth persists only because nobody has tried to identify what those jokes are, but the sentiment has some truth. Even as times change, humor stays constant, which is why we can still appreciate many jokes dating back to Roman times: "A garrulous barber once asked his client how he should cut his hair," goes one gag shared more than two thousand years ago. "'Quietly,' the client replied." It may be that traditional jokes are rare, if not dead, and that humor is best understood not through one-liners but in terms of conflicting thoughts and feelings. Yet it's still useful to analyze jokes because there's no better way of understanding how humor affects us all differently. There's something universal about humor, despite its many forms. What better way to recognize different humor types than to see them in action in the form of jokes?

Probably the most successful attempt at categorizing humor types has the least funny name possible: the 3WD Humor Test (WD stands for *Witz Dimensionen,* or "joke dimension"). It was developed by German researcher Willibald Ruch, who asked subjects a series of questions about jokes and cartoons and, based on these judgments, grouped their humor preferences into three types. The first type is called incongruity-resolution, which typically involves violating expectations in novel ways, with punch lines leading to surprise or relief. The second is called nonsense humor, which is funny only because it makes no sense at all. The third is sexual humor, which is frequently offensive or possibly taboo. Though the content of individual jokes varies, Ruch showed that the way they provoke us generally falls into one of these three categories, with the most popular jokes relying a little on each.

Another approach is to rank statements into categories depending on how well they describe our humor tastes. This is the technique used by the Humorous Behavior Q-Sort Deck, which involves one hundred cards containing printed statements ranging from simple (e.g., "Is sarcastic") to reflective (e.g., "Only with difficulty can laugh at personal feelings"). Participants sort the cards into nine decks, depending on the personal relevancy of the statements, and their sense of humor is assessed in terms of how social, restrained, or cruel it is. Extensive research using this test has revealed that American tastes in humor tend to be socially warm and reflective whereas British humor leans more toward the spirited and amusingly awkward.

But trying to measure humor without considering the psychological background of subjects is difficult, because we can't see where their conflicts lie. We're forced to make our best guess—and though doing so may prove useful, it can be tricky too. Perhaps this is why one scientist, Richard Wiseman, decided to stop asking subjects to characterize jokes altogether. Instead, he simply asked them a single question: "Is this joke funny?" He didn't ask them why, and he didn't make them visit his lab either. Rather, he enlisted help from the British Association for the Advancement of Science and started a website. One year and 1.5 million responses later, he stumbled upon the funniest joke in the world.

Wiseman is a psychologist from the University of Hertfordshire just north of London. He has written four books and is generally considered one of the most influential scientists in Britain. Though not a humor researcher by training, he's had plenty of experience exploring unusual topics such as deception, the paranormal, and self-help. He also is credited by Guinness as the lead researcher on one of the largest scientific experiments of all time.

His project, called LaughLab, started with a simple question: What makes a joke funny? To research this question, he asked people to answer a few short questions about themselves, then to rate the funniness of a random sample of jokes, based on a "giggle-o-meter" scale of 1 to 5. Since he also wanted to keep a fresh supply of jokes, he added a section where people could submit their own personal favorites. Thanks to some free publicity and plenty of international interest, millions of people flocked to his website. In all, Wiseman received over forty thousand jokes, many of which had to be rejected because they were too vulgar to be shared with a wider audience. Wiseman included jokes that he didn't think were particularly funny, in case he accidentally missed the humor. For example, the joke *What's brown and sticky? A stick* was submitted more than three hundred times, and Wiseman left it in, because he figured that such a large number of people must know something he didn't.

In addition to telling him what jokes people found most funny, the experiment produced vast amounts of information, thus allowing for some very specific analyses. For example, Wiseman found that the funniest jokes were 103 letters long. This particular number wasn't special; there just had to be some length where ratings peaked, and 103 letters was it. Since many of the jokes included references to animals, he was also able to identify what animal was the funniest. Ducks, interestingly, won that prize. Maybe it's the webbed feet, Wiseman mused, but if a joke-teller has the option of giving the starring role to a talking horse or a talking duck, the choice is clear. The funniest time of day: 6:03 in the evening. The funniest day: the fifteenth of the month. Wiseman's data yielded an almost endless supply of discoveries.

One of the most interesting findings was how humor varies based on nationality. Germans thought every joke was funny. Scandinavians ranked closer to the middle, and also had the unfortunate tendency to include the words "ha ha" at the end of their entries, as if reassuring the reader that they had just experienced a joke. Americans showed a distinctive affinity for jokes that included insults or vague threats.

Here's a joke particularly liked by Americans, less so by others:

> TEXAN: *"Where are you from?"*
>
> HARVARD GRADUATE: *"I come from a place where we do not end sentences with prepositions.*
>
> TEXAN: *"Okay—where are you from, jackass?"*

Europeans, in turn, showed an affinity for jokes that were absurd or surreal. Here are two more examples:

> *A patient says: "Doctor, last night I made a Freudian slip. I was having dinner with my mother-in-law and wanted to say, 'Could you please*

*pass the butter.' But instead I said, 'You silly cow, you have completely ruined my life.'"*

*A German Shepherd went to the telegram office, took out a blank form, and wrote: "Woof. Woof. Woof. Woof. Woof. Woof. Woof. Woof. Woof."*

*The clerk examined the paper and politely told the dog: "There are only nine words here. You could send another 'Woof' for the same price."*

*"But," the dog replied, "that would make no sense at all."*

British people's taste for the absurd is informative, and also well supported by separate laboratory work. From sense-of-humor questionnaires, we know that the British consistently express a preference for deadpan or irreverent humor, just as Americans enjoy teasing and kidding. What else would you expect from a country that gave us this zinger: *Mommy, what do you call a delinquent child? Shut up and hand me the crowbar!*

Wiseman also discovered that humor varies widely based on gender. Women responders to his website distinguished themselves not in terms of their favorite jokes but in terms of what jokes they rated lowest. For example, although male responders consistently rated put-down humor highly, women seldom agreed, especially when the targets were women. We'll discuss this issue more later, but for now, in the interests of science, let's look at a joke that more than half the men enjoyed but only 15 percent of the women rated positively:

*A man driving on a highway is pulled over by a police officer. The officer asks: "Did you know your wife and children fell out of your car a mile back?" A smile creeps onto the man's face and he exclaims: "Thank God! I thought I was going deaf!"*

Future chapters will examine what makes each of these jokes funny, but a general observation can still be made here—each is short, just under half the "maximum funniness" length of 103 letters. The comedy writer Brent Forrester refers to this preference for brevity as the Humor and Duration Principle, otherwise known as "The shorter the better." He even gave it a formula: $F = J/T$. If $F$ represents the level of funniness, then funniness depends on both the quality of the joke, $J$, and the amount of time needed to tell it, $T$. The best jokes are always lean. No fat, no extra words.

Wiseman's study did have its shortcomings too. For example, only English-speaking people were able to participate, and the funniest jokes didn't always succeed. (That isn't just my opinion; it's Wiseman's too.) That's because the jokes that avoided extremes, the "safe" jokes, tended to receive the most votes, leading to an unfortunate tendency toward mediocrity. This shouldn't be surprising since we've already learned that humor is by nature confrontational—sometimes cognitively, sometimes emotionally, and sometimes both. Because people vary in terms of how much they like to be provoked by their jokes, the most popular jokes tend to cluster near, but still below, the most typical "provocation threshold." Too high, and some people laugh wildly but others not at all. Too low, and nobody is impressed.

Fortunately, Wiseman was pleased with the eventual winner, if only because it just barely beat out the second-place competitor. The latter wasn't a bad joke, per se; it just wasn't all that good and most people have heard it too many times already. (The punch line reads "Watson, you idiot, it means that someone stole our tent," in case you want to look it up.) Wiseman frequently tells both of these jokes in front of audiences because his research is often featured on television and at conferences, and most of the time neither one gets a laugh. One problem is the jokes themselves, for sure. But another is the delivery. Like most humor researchers, Wiseman has no comedic training, and so by his own admission he doesn't know how to tell a joke. That's another big issue in humor research, one that will get plenty of attention in Chapter 7.

What was the winner? Don't say I didn't warn you.

*Two hunters from New Jersey are out in the woods when one of them collapses. He doesn't seem to be breathing and his eyes are glazed over. The other guy whips out his phone and calls the emergency service. He gasps, "I think my friend is dead! What should I do?" The operator says, "Calm down. I can help. First, make sure he's dead." There is a silence, then a shot is heard. Back on the phone, the guy says, "Okay, now what?"*

# 2

## ≫ THE KICK OF THE DISCOVERY

*The prize is the pleasure of finding the thing out,*
*the kick of the discovery.*
                                        —RICHARD FEYNMAN

TO INTRODUCE THE IMPORTANT ROLE OF DISCOVERY IN HUMOR, let's look at a 2008 experiment conducted at Northwestern University. In contrast to our previously described studies, this one had nothing to do with humor. Instead, the scientists asked subjects to solve problems notorious for being exceptionally hard. So hard, in fact, that they couldn't be solved analytically. The problems required what scientists call *insight*. Insight is what happens when we have no idea how to solve a problem and, instead, must rely on answers that pop into our heads for no apparent reason.

Another distinctive aspect of the Northwestern study is that a huge, multi-ton magnet surrounded subjects' heads as they worked, altering the spin of protons in their brains so that scientists could tell what parts were most active.

The machine was an MRI, which allowed the subjects' brain activity to be viewed as they followed the experimenters' instructions. Three words were shown at a time, and though these words were unrelated, each was closely associated with another common word that wasn't shown. The task was to guess that fourth word. As soon as subjects had an answer, they pressed a button, and for each set of words they were given fifteen seconds to identify a solution before the next three words appeared. Let's see an example so you can try it yourself:

> *tooth*
>
> *potato*
>
> *heart*

Obviously, the task isn't easy. For most people, the first word that comes to mind after reading *tooth* is *ache.* That fits with *heart,* but not with *potato.* The first word most people associate with *potato* is *peel,* but that doesn't fit with either of the other two. You can see why this is called an insight problem. Brute force analysis doesn't work. Let's consider another example:

> *wet*
>
> *law*
>
> *business*

This time, let your mind relax. Even if the answer feels close, don't let your brain slip into an analytical mindset. Ignore any similarities you might perceive between the words *law* and *business,* because those will hold you back. The only way you'll come up with an answer is to let your mind go blank. Here's one last example:

> *cottage*
>
> *Swiss*
>
> *cake*

This last triad is easier, and hopefully you came up with the answer *cheese*—just as you may have come up with *sweet* for the first example and *suit* for the second. The task is called the Remote Semantic Associates, and it's known for being exceptionally difficult. So difficult, in fact, that a study involving hundreds of people found that fewer than 20 percent were able to solve either of the first two problems within fifteen seconds. When given thirty seconds, most were able to solve the second one. And the last one, the one with the solution *cheese* (the easiest of the nearly 150 original problems), was solved by 96 percent of the subjects, most in two seconds or less.

Human insight is an amazing thing, and it's especially important for humor, as we'll soon see. Some connections between insight and humor may already be apparent, such as the close link they both share with pleasure. We enjoy coming up with solutions, whether in the form of punch lines or insight problems like the ones above. That's what physicist Richard Feynman meant when he described "the kick of the discovery." His greatest award wasn't the Nobel Prize, he claimed, but the pleasure of having a job that involved discovering new things. We take pride and pleasure in solving problems because our brains are programmed with an inherent desire to explain. According to Berkeley psychologist Alison Gopnik, this drive is as fundamental as our desire for sex. "Explanation is to cognition as orgasm is to reproduction," she says. Thinking without understanding is as unsatisfying as sex without . . . well, you know.

Consider, for example, a study conducted by the psychologist Sascha Topolinski at the University of Wuerzburg in Germany. Topolinski showed his subjects word triads similar to our earlier examples, except that he included sets with no solution at all (for example, *dream, ball,* and *book*—good luck!). Rather than monitoring subjects' brains using an MRI, he closely examined their facial muscles, looking for responses that might give indications of their thought processes. Without informing his subjects that some word sets had shared associates, he found that triads sharing a single word in common elicited a very interesting reaction. Specifically, when subjects read those triads, the muscles responsible for smiling and laughing (zygomaticus major muscles) were activated and the muscles responsible for frowning (corrugator supercilii) were relaxed. In other words, although the subjects thought they were simply reading unrelated words and didn't even try to come up with solutions, they responded as if they had just heard a joke. They experienced pleasure.

Perhaps this is why Karuna Subramaniam, the Northwestern University scientist who conducted the experiment described in this chapter's opening, also had subjects rate their mood before starting. Though mood can be a difficult thing to measure, scientists have developed several tests—such as the Positive and Negative Affect Schedule and the State-Trait Anxiety Inventory—to identify the degree to which people feel positive or anxious at any given moment. By assessing her subjects' emotions at the point they entered the lab, Subramaniam was able to determine whether mood had any effect on how well they solved the insight problems. It did. Subjects in a good mood not only solved more problems than those in a bad mood, they also engaged a specific part of the brain responsible for managing conflict. That region is called the anterior cingulate.

In this chapter we'll take a closer look at humor by examining the three stages our brains go through when transforming ambiguity and confusion into pleasure. Along the way we'll see how these stages allow us to both understand jokes and solve problems using insight. We'll also see how one region of the brain, the anterior cingulate, plays a special role in keeping the rest of our minds in check.

## THE THREE STAGES

Interpreting our world is a creative event. We are by nature hypothesis-generating creatures, meaning that we don't just passively take in our environment but, instead, are always guessing what we need to do or say. Sometimes these guesses are wrong, which isn't a bad thing. It's good, because detecting errors is how our brains turn conflict into reward. Reward comes in the form of pleasure-inducing neurotransmitters, such as dopamine, that are released only when the conflict is resolved. Without this conflict, there would be no way to regulate reward, and so everything would give us equal amounts of pleasure. And, as the saying goes, if everything makes us happy, then nothing does.

These three stages, which I call *constructing, reckoning,* and *resolving,* are key not only for humor but for all aspects of complex thinking. When solving insight problems, we must generate possible solutions while also inhibiting "false alarms" and other incorrect answers. This process introduces the potential for lots of conflict, and it takes our brains several steps to work through the challenge. Let's look at these steps individually to see why each is so important.

### Constructing and the Anterior Cingulate

Why are insight problems so difficult? Is it because we have too many words floating in our heads to make sense of them all? Absolutely not. The challenge of insight problems is that our minds get stuck on wrong answers. We have trouble coming up with the correct response because the wrong ones keep pushing themselves upon us.

The initial triad we looked at—*tooth, potato, heart*—is a good example. For each word, the solution *sweet* isn't the first one most people think of. It isn't even in the top ten. We know because there are databases containing what scientists call semantic associates—words that come to mind when subjects are presented with a prime such as *tooth*—and *sweet* appears near the bottom of the list for all three of our triad words. In fact, even as I write this chapter, and though I've seen the answer many times, the word *ache* still keeps coming falsely to mind. During an early draft of this book, I even misidentified it as the correct solution. Thank goodness for proofreaders.

I call the first stage of the humor process *constructing* to show how active we are in processing our environment. When solving problems, we don't simply search our memories for possible solutions. Rather, we let our brains go to work generating lots of possible answers, some of them useful (*sweet*) and others not (*ache*). We do the same when reading jokes—though, in this case, misdirection comes before the punch line. *One morning I shot an elephant in my pajamas. How he got in my pajamas, I don't know* goes the classic Groucho Marx joke. Who wears the pajamas depends on how far along you've read.

The brain is a complex beast. There are separate regions for vision, hearing, and language, plus several guiding our movements. There are regions that activate when we do complex math, others that store new memories, and still others that help us recognize faces. The only thing more amazing than the brain's specialization is that it does so many things that evolution could never have fully prepared it to do. So, it shouldn't be surprising that when the brain gets to thinking about things, it makes some wrong turns.

Consider the fact that your brain has between 10 billion and 100 billion neurons. That number is so large it's meaningless, so let's compare it to the number of people living on Earth, which is roughly 7 billion and growing. That's pretty close to the lower estimate for the size of your brain, so let's consider these two systems further. Imagine that at this very instant the entire population of the United States decided to scream. That would be comparable to the neural activity in your brain—at rest. A brain not at rest would be ten or a hundred times more active, so now it's not just America screaming but all of Asia too. All it takes is for multiple parts of your brain to disagree, and pretty soon you've started World War III.

The brain manages this complexity the same way humans do—by building "governments." As I've noted, its various regions are specialized for nearly everything we do, and although nobody knows exactly how many specialized modules the brain has, it's probably close to the number of governments in the world. There has to be a way to manage all these voices, and for humans the solution is to create entities like the United Nations. The UN isn't itself a government, as it has no land, economy, or political goals. It simply keeps an eye on everyone around it, hearing complaints and keeping troublemakers in check. The brain has a UN, too. It's called the anterior cingulate.

Located near the center of our brain, just above the corpus callosum connecting the two cerebral hemispheres, the anterior cingulate is in a

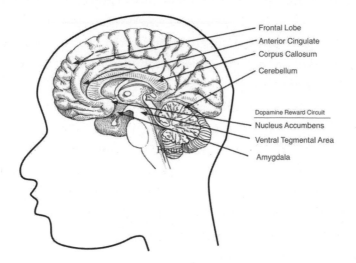

Frontal Lobe
Anterior Cingulate
Corpus Callosum
Cerebellum

Dopamine Reward Circuit
Nucleus Accumbens
Ventral Tegmental Area
Amygdala

FIGURE 2.1. Selected regions of the human brain.

perfect position to oversee the rest of the brain (see Figure 2.1). In front is the frontal lobe, our primary reasoning center and the region responsible for initiating movement. Behind are the parietal and temporal lobes, which help with reasoning, as well as language and memory. And as part of the brain's limbic region, the anterior cingulate is closely connected with the amygdala, the nucleus accumbens, and the ventral tegmental area—regions that, as noted earlier, are key to the dopamine reward circuit.

We know that the anterior cingulate is especially important for insight because we can observe activity in subjects' brains prior to solving problems like our word triads. Most parts of the brain become less active as subjects prepare to solve these difficult problems, but the anterior cingulate is different. It becomes *more* active, because rather than coming up with solutions, it handles conflict. The Remote Semantic Associates task doesn't appear at first to be one driven by conflict, but it is. As we already discussed, the solution is seldom the first one anybody thinks of. Coming up with a solution requires "holding back" more potent responses. The part of the brain that thinks it has the easy answer needs to be "shut up" so that softer voices can be heard. And telling others to shut up is exactly what the anterior cingulate is good at.

A good way to understand the anterior cingulate is to explore the Stroop task, named for John Ridley Stroop, who developed it in 1935. He found that when we are asked to identify the color of something, we are slower and less accurate when that thing is a color word. For example, it's easy to identify the color of four asterisks printed in red, but much harder to provide a correct response when the items printed in red are the letters *B-L-U-E*. Why? Because now we have two competing responses. The human mind naturally wants to read, and preventing it from doing so is almost impossible. If you don't believe me, try performing this simple experiment at home: watch an English-language movie tonight with the subtitles showing. I guarantee that you'll be reading every word on the screen, even though you understand exactly what is being said.

What does this have to do with the anterior cingulate? Well, the Stroop task is exactly the kind of thing the anterior cingulate is specialized for, because it's the only brain structure able to keep the reading regions silent so that the color-identifying regions can respond. And it's especially effective at managing such control when we're in a good mood, which is why the Stroop effect disappears when we're happy. When subjects are asked to recall positive life events such as vacations or birthdays immediately before performing a Stroop task, they no longer have difficulty ignoring the conflicting color words. Just as insight is positively correlated with happiness, happy people are better at maintaining focus while identifying the color of fonts.

Mood and happiness are also important to *constructing*, as we'll soon see. As the brain constantly debates over what to say or do, the anterior cingulate stays very busy, and so anything that provides help can have a big influence. Positive mood improves focus by helping the anterior cingulate hold back unwanted responses, such as *ache* in the insight problem and *B-L-U-E* in the Stroop task when the font is red. If the anterior cingulate is like the UN, then positive mood is its operating budget.

However, it's important to realize that we are not passive actors in our environment. We don't just take in information, we create it. We are constantly developing theories and expectations about our surroundings, then revising them when necessary. This phenomenon is observed not just in laboratory settings but on the scale of whole societies, past as well as present. Our ancestors interpreted lightning as anger from the gods, and eclipses as dragons eating the sun. Such beliefs invaded science too. Aristotle, who was born almost two thousand years before the invention of the microscope, thought that life spontaneously arose from slime and mud exposed to sunlight, because this was his only explanation for mold. And Isaac Newton, who lived during an age when chemistry offered no explanation for the mysterious existence of gold, wrote more than a million words about the subtleties of alchemy.

To show how ubiquitous our need is to construct such theories, and also to see its link with humor, let's consider one last study before moving on to our second stage, *reckoning.* This study was conducted by the Swedish psychologist Göran Nerhardt, who wanted to know if he could induce laughter using materials that weren't funny at all. He didn't even tell his subjects that they would be participating in a humor study. Instead, he simply instructed them to pick up a series of objects and waited to see just how far their false expectations could take them.

Nerhardt's task was quite straightforward. Subjects picked up objects of various weights (e.g., between 740 to 2,700 grams, roughly 2 to 6 pounds). Then they were asked to classify each on a 6-point scale, ranging from very light to very heavy. This sequence was repeated several times, after which subjects were given an object that was much lighter than the others—just a tenth of a pound. They weren't told anything special about this last object. They were simply asked to make a series of judgments about items that weren't funny at all.

Yet Nerhardt found that, when the subjects were asked to make a judgment about the final, wildly incongruous weight, most of them laughed despite being given no indication that it was intended as a joke. Not only that, but the more the weight differed from those lifted earlier, the more they laughed at this absurdly light one.

In the forty-plus years since this experiment was first conducted, the design has been varied several times, and each time the subjects' reaction is the same—they find the last, incongruous weight funny. There's nothing humorous about the weights themselves. The subjects simply have to construct an expectation. And when that expectation proves false, they have no other choice but to laugh.

### Reckoning in a Confusing World

Now that we've explored the concept of *constructing,* let's see what our brains do with all these wild expectations. Only by observing the consequences of our false starts can we understand why they so often lead to humor. This means familiarizing ourselves with *reckoning,* the jettisoning of our mistakes so that we can uncover new interpretations.

My guess is that if you asked a hundred experts what the key ingredient of humor is, most would say surprise. Surprise is special because it affects us in so many different ways. It's what makes insight problems unique, because for these tasks we have no idea how close we are to a solution until we already have it. That's what defines insight problems. Research by Janet Metcalfe at Indiana University showed that confidence in being close to an answer for insight problems is *inversely* related to actual progress. In other words, the closer we think we are to a solution, the farther away we really are. Surprise isn't a by-product of completing these tasks, it's a *requirement.*

Surprise is important for humor the same way it's important for insight—we take pleasure in being pulled from false assumptions. Punch lines catch us by surprise, and the more we set our expectations on one interpretation, the more we allow ourselves to be caught off-balance by the actual turn of a joke. A joke that you've heard before isn't inherently less funny. It's just old news, and so it no longer gives you surprise. An insight problem that you've seen before isn't fun or challenging either, because you no longer need insight to solve it. You just need a bit of memory.

*Reckoning* is the process of reevaluating these misperceptions, usually leading to a pleasant surprise. We enjoy discovering our mistakes because surprise is one of our most valued emotions, as fundamental as happiness or pride. Scientists have even quantified the importance of surprise by asking people about recent emotional experiences. This is what Craig Smith of Stanford University did when he asked subjects literally thousands of questions regarding recent events in their lives, questions like "How pleasant or unpleasant was it to be in this situation?" and "When you were feeling happy, to what extent did you feel that you needed to exert yourself to deal with this situation?" Using advanced data analysis, he was able to locate the subjects' emotions along certain dimensions, including pleasantness and the amount of effort they required from the person experiencing them. Figure 2.2 shows how surprise ranked, compared to other emotions.

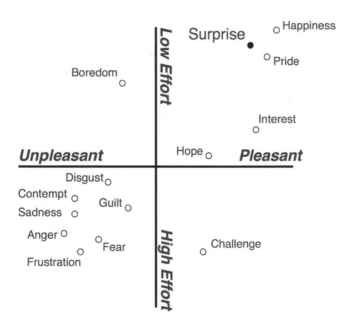

FIGURE 2.2. Emotions as they vary by pleasantness and effort involved in their experience. Adapted from Craig Smith and Phoebe Ellsworth, "Patterns of Cognitive Appraisal in Emotion," *Journal of Personality and Social Psychology* 48 (1985): 813–838. Published by the American Psychological Association. Adapted with permission.

As it turns out, surprise holds a special place, near the top of the diagram. Since the axes measure pleasantness and effort required for their experience, this means that surprise is one of the most positive and natural emotions we experience.

Surprise leads to pleasure in lots of contexts, not just humor. German psychologist and art theorist Rudolf Arnheim presented perhaps the most graceful example of pleasant surprises when he analyzed, of all things, a violin sonata by the Baroque composer Jean-Marie Leclair. Leclair, who wrote nearly a hundred major works in the mid-eighteenth century, was well known for creating sophisticated, cerebral violin concertos. In one of his last works, there's a point near the middle where he suddenly includes a note that is harshly out of key. At first it sounds dissonant, and the listener wonders if perhaps there has been a mistake. But the same note occurs again, and then another surprising note, and soon we realize that the composer has switched keys in the middle of the performance. An examination of the music in written form reveals that the change is entirely intentional—a note written as B flat is identified as A sharp later in the same measure, conveying Leclair's message that it serves different purposes for the old and new keys. In just a few notes the listener is compelled to discard previously held assumptions about the piece and to listen to it in an entirely new way. And the experience is richer for it.

Arnheim explains that such sudden shifts occur in architecture too. Take, for example, the Hôtel Matignon, the Paris mansion designed in 1725 by architect Jean Courtonne that now serves as the home of the French prime minister, Jean-Marc Ayrault. At the time it was built, tradition dictated that buildings be built symmetrically about an axis connecting the front and rear entrances. But this was impossible for the Hôtel Matignon given the surrounding streets, so the architect did the only thing he could—he shifted this axis inside the building itself. Visitors entering either entrance see everything laid out in the expected, symmetrical fashion. But further on, there's a point where everything suddenly shifts and they're off-center relative to the entrance they used, and are now centered about the opposite one. Some call it cheating, others call it brilliance, but everyone appreciates that this shift is what makes the building so pleasurable to live in—including its current resident.

These phenomena have an equivalent in the realm of humor, and it's called paraprosdokia. Paraprosdokia is speech that involves a sudden and surprising shift in reference, usually for comedic effect. Take, for example, the following quote by Stephen Colbert: "If I am reading this graph correctly, I'd be very surprised." Colbert was looking at polling data for the 2008 presidential elections—data that under even the best of circumstances would be difficult to interpret. At first it sounds like he's preparing an insightful and cutting remark. Instead, we realize he's basking in the ignorance we all feel when trying to interpret such numbers. The joke required no setup or punch line. All it needed was for the listener to "jump the gun" regarding what Colbert was actually saying.

Not surprisingly, the brain region responsible for catching these false starts is the anterior cingulate. We know this from studies like the one conducted by biologist Karli Watson of the California Institute of Technology, who wanted to see if any particular brain region was especially important for surprise. To do that, she showed subjects cartoons while they were monitored using an MRI scanner, and (as in previous studies) she made sure that some cartoons were funny whereas others were not. As an additional manipulation, she varied the nature of the cartoons so that some relied on sight gags whereas others depended on captions and language. Variations like this can have big impacts on how the brain responds, since visual centers are very different from language ones—so she expected the jokes to enlist entirely different regions. But were any regions activated in common?

The answer, of course, was yes. Both the dopamine centers and the anterior cingulate were active for each kind of joke. Not only that, but the funnier the jokes, the more engaged was each subject's anterior cingulate.

Studies like this provide a great example of *reckoning* because they show that what elicits laughter isn't the content of the joke but the way our brain works through the conflict the joke elicits. This can be seen in Colbert's quip as well as in Leclair's violin sonata and Courtonne's Hôtel Matignon. We take joy in recognizing our mistakes. Though we often think of punch lines as involving misdirection, it's actually our anxious brains that supply the false interpretations. There were no dissonant notes in Leclair's sonata, just as there was no actual contradiction in Colbert's one-liner. The enjoyment of both comes solely from overriding a false expectation created within ourselves. In this way, *reckoning* builds on *constructing* by forcing us to reexamine false expectations.

To see how all this eventually turns into a joke, let's finally explore the concept of *resolving*.

### Resolving with Scripts

> *A large woman sits down at a lunch counter and orders a whole fruitcake. "Shall I cut it into four or eight pieces?" asks the waitress.*
> *"Don't cut it," replies the woman. "I'm on a diet."*

Is this joke funny? Unless you have a special affinity for fruitcake humor, your answer is probably no. But at first glance it seems like it should be, because the woman's response is definitely surprising. It's so surprising that it makes no sense at all. Consider, then, this alternate ending:

> *A large woman sits down at a lunch counter and orders a whole fruitcake. "Shall I cut it into four or eight pieces?" asks the waitress.*
> *"Four," replies the woman. "I'm on a diet."*

Now is it funny? Again, you probably didn't laugh out loud, but I bet you at least found it funnier than the first version. The reason is that this second version provides an explanation for the sudden shift in perspective. It isn't enough just to introduce surprise in a joke; we must also provide a shift in perspective. I call this third stage of the humor process *resolving*.

When studying humor, we need a way to characterize the expected and actual outcomes of a joke. For our fruitcake story, we see there are several words signaling an expectation of gluttony. There's the fact that the woman orders a whole fruitcake, not just a slice. She's also described as large. All this background suggests that she's really looking forward to the cake. When she asks for four slices instead of eight, one interpretation—the one influenced by her weight—is that she thinks four slices means fewer calories. The other interpretation, the right one, is that strokes of the knife have nothing to do with calories or the amount of cake.

Pretty tedious, huh? After such an analysis, it's clear why dissecting humor is often likened to analyzing a spider's web in terms of geometry. It loses its grace.

I apologize for breaking down such a bland joke, and I promise not to do it again. But it's important to recognize that joke construction is complicated. To compare contrasting meanings, we need a scientific way to characterize all the false assumptions involved in the joke. We need a way to measure distances between intended and unintended meanings to get an idea how funny a joke can be. And, perhaps most importantly, we need to understand why people laugh at some incongruities—such as a woman thinking four large slices of fruitcake are healthier than eight small ones—when much bigger incongruities—such as a woman walking into a diner and ordering an entire fruitcake—are seemingly ignored. To do that, we need to understand scripts.

After graduating from the University of California with a PhD in psychology, I initially worked as a postdoctoral researcher with computer scientist and neurologist James Reggia. I was excited to work with Reggia because he was interested in nearly everything. He studied not only hemispheric laterality (my own specialty) but also language and memory. He specialized in artificial intelligence and chaotic swarming, an emerging field that uses artificial life to examine large-scale problem spaces. He even taught classes on machine evolution and expert systems. In short, he was the kind of person who knew something about nearly everything. So, when we first met in a restaurant in Columbia, Maryland, his first words to me were a surprise.

"I'm looking forward to working with you. I've never worked with a boxologist before."

Though I had no idea what he meant, when he explained I not only understood but agreed with Reggia's characterization, and we've become close friends. Reggia meant that we psychologists, by nature, love drawing boxes. We take complex cognitive and social phenomena, and to understand them we break components into processes and surround them with boxes. We draw arrows between the boxes to show how they influence each other, and when we get especially spirited we take the boxes away to make more room, leaving only words and arrows. It can seem silly at times, but often we have no choice because what we study is complex. Which is why I would like to direct you to the joke in Figure 2.3 and let you see just how boxy our analyses can be.

I'm guessing that, again, you didn't laugh. If you did, then you should stop reading now because I have nothing more to offer you. Now, how about if I present the joke in a format you're more accustomed to?

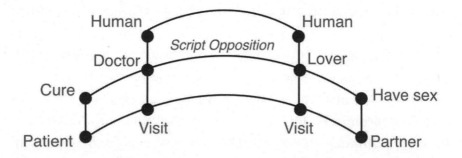

FIGURE 2.3. Graphic representation of the joke starting *"Is the doctor home?"* Get it? Adapted from *Humor, the International Journal of Humor Research.* Band 15, Heft 1, Seiten 3–46, ISSN (Online) 1613–3722, ISSN (Print) 0933–1719, DOI: 10.1515/humr.2002.004, De Gruyter Berlin/Boston, January 2006.

*"Is the doctor home?" the patient asked in a bronchial whisper.*

*"No," the doctor's young and pretty wife whispered in reply. "Come right in."*

Though this second presentation looks starkly different from the first, the joke is the same in both cases. The first is just a graphic representation of all the joke's key elements, as identified by Salvatore Attardo. A linguist at Texas A&M University, Attardo is one of the most prolific humor researchers in the world. His primary contribution to the field is what he calls the General Theory of Verbal Humor (GTVH), which explains how jokes are actually just different ways of manipulating *scripts*. To understand what that means, we need to take a closer look at Figure 2.3 and see how it maps onto the joke format we're more familiar with.

Let's start by taking a close look at the words. Each word reflects a different script, which is a chunk of information describing some object, action, or belief. Scripts are different for everybody, and there are no rules for what must be contained in a person's script. For me, the script for *doctor* includes the fact that he or she sees patients, prescribes medicine, and probably plays golf. Depending on your own exposure to specialists such as pediatricians or psychiatrists, your script might include other expectations such as lollipops and offices with couches. Babies are born without scripts. Scripts must be learned.

Scientists use scripts to study humor because they allow for systematic analysis, as we see in Figure 2.3. Note that the left side of the figure represents all the scripts that are activated by the initial interpretation of the joke. We initially think the patient is going to the doctor to seek a cure. (The circular "nodes" represent scripts, and the lines represent meaningful connections between them.) Then, when the doctor's wife invites him inside, we see that several scripts have been falsely activated. The patient is not sick. He's not seeking a doctor either, he's soliciting his lover. From the figure, you see that the common element between doctor and patient is *cure*. The corresponding link between lover and partner is quite different.

The idea of scripts is an old one, rooted in almost fifty years of psychological research. "Scripts were really meant to be an umbrella term for all the knowledge humans have to describe their world," says Attardo. "When Victor [Victor Raskin, original developer of Script Theory] introduced the idea, he intended it to be general. In the 1970s, there was a proliferation of research on things like schema, frames, schemata—all slightly different ways of describing how humans organize information. Some were defined more formally or operationally. But they were all trying to do the same thing. They were trying to say how people manipulate knowledge about their world."

All of this analysis may seem rather technical for a sub-par joke, but it does illustrate some important requirements for humor. First, in order to succeed, a joke must activate multiple scripts. Second, those scripts must oppose each other—and the greater the opposition, the funnier the joke. The key opposition here is between *cure* and *having sex.* Being treated for asthma or tuberculosis is about as different from having an afternoon tryst as you can get.

Another benefit of thinking about humor in terms of scripts is that it allows for certain incongruities to be highlighted, and others ignored. Consider the following joke to see what I mean:

> *A bear walks into a bar and approaches the bartender. "A martini . . . dry."*
> *The bartender asks: "What's with the pause?"*
> *"I don't know," the bear replies. "I was born with them."*

This joke relies on a *pun,* which is essentially a conflict between scripts based on phonological ambiguity, but that's not the point. The point is—what's a bear walking into a bar for? Why didn't the bartender run for his life? And how could a bear possibly hold a martini glass, anyway? We ignore these incongruities because we quickly recognize they aren't part of the joke. The key script opposition is between the words *pause* and *paws,* which has nothing to do with the bear's sudden ability to talk and consume gin. Some humor actually exploits these seemingly ignored incongruities, as seen in the long his-tory of elephant jokes:

*How did the elephant hide in a cherry tree? He painted his toenails red.*

I personally love elephant jokes, not because they're clever but be-cause they make fun of the very concept of surprise. The script oppo-sition at the heart of this particular joke is the least salient aspect of a tree-climbing elephant. Never mind that any tree would surely break under elephantine pressure, or that pachyderms don't have opposable thumbs for climbing, much less fingers. What I really want to know is what color were its toes! The whole joke is absurd because its most highly activated script, elephants, conjures up thoughts of size and weight, thereby undermining every other aspect of the joke.

Research has found that such background incongruities aren't just tolerated, they make jokes funnier. I'm referring to a series of experi-ments by the psychologist Andrea Samson at Stanford, who instructed subjects to view cartoons that either included background incongruities or omitted them entirely. For purposes of experimental control, two versions of each cartoon were used: an "extra-incongruity" one and a realistic one. Subjects saw mixes of each and were tasked with rating how funny they thought each was. For example, one cartoon showed a mother and father penguin standing in the Antarctic wilderness, cel-ebrating with wild gesticulations: "He just spoke his first word," one says. "Great! What is it? 'Mama'? 'Papa'?" says the other. The second panel shows both penguins standing next to their offspring, who is ex-claiming: "Damned cold!" In the realistic version, the words remained the same but penguins were replaced with Eskimos.

Samson found that subjects preferred the jokes with background incongruities. Penguins, apparently, made the jokes funnier.

To see what all this has to do with *resolving,* let's take one final look at the brain, this time using an electroencephalogram, commonly known as the EEG. The experiment was conducted by psychologist Peter Derks for a 1991 conference held by the International Humor Society in

Ontario, Canada, and it involved the measurement of subjects' brain activity using electrodes placed strategically along their scalps. The electrodes couldn't tell what the subjects were thinking, but they did show when their brains got particularly busy. While hooked up to these electrodes, twenty subjects read a series of jokes, each ending in a final word providing a surprising punch line. At the same time, Derks and his colleagues monitored the zygomatic muscles controlling the subjects' mouths, a useful method for scientifically determining whether someone has laughed or smiled.

When the EEG data were analyzed, Derks saw that the subjects produced two very different electrophysiological responses to the jokes. The first was a peak in activity called the P300. This occurred about a third of a second after the last punch line word and took the form of a sudden, positive spike in electrical activity. In short, the subjects' brains got very busy pretty soon after the jokes were completed. Then, about a hundred milliseconds after that, the EEG showed an N400—a negative deflection also representing a sudden increase in electrical activity, again due to increased brain processing.

A couple of things about the EEG are important to note here. First, the positive or negative nature of any observed EEG effect is meaningless because it depends on the way neurons are oriented in the brain, which has nothing to do with how we think. Second, and more importantly, the timing and identity of the observed electrical potential mean everything. In fact, the P300 effect has been observed in hundreds of studies, if not thousands. From these studies, scientists have learned that it always reflects an orienting response. When people see something they don't expect, or something that grabs their attention, they invariably show a P300. The N400 is just as well studied, though it reflects a different kind of processing. The N400 occurs when the brain has to fit new information into existing knowledge—a process called semantic integration.

Sadly, the presence or absence of P300 and N400 effects alone tell us nothing about how subjects process jokes, but when they are combined with muscular-response data collected from subjects' faces, a new picture emerges. Derks found that subjects clearly perceived some jokes as funnier than others, indicated by the subjects' zygomatic muscles. When Derks separated the trials containing jokes that were funny from those that weren't, he saw that all of the subjects showed a P300 effect, regardless of joke funniness. However, the N400 effect emerged only when the subjects' zygomatic muscles were activated. In other words, jokes that weren't funny didn't make people laugh, and didn't elicit semantic integration or an N400.

Derks had found proof that humor involves more than just being shocked or surprised. Jokes that weren't funny still brought on an orienting response—a P300—because they included a surprising punch line. But that's all they did. They didn't lead to a satisfying resolution, and thus never made it to our third stage of humor processing. They didn't activate an opposing script, allowing for the joke to "come together." And so, after encountering the incongruity, subjects' brains became silent.

Derks's findings clearly differentiate between *reckoning* and *resolving* because they show that it's one thing to hear a joke but quite another for that joke to make us feel satisfied. Putting everything together and "getting" the joke is distinctly separate from being shocked or surprised, and I call this stage *resolving* because humor requires not just dealing with the unexpected but activating a new frame of reference.

Interestingly, the anterior cingulate has been closely linked with the P300 but is unrelated to the N400. In other words, the anterior cingulate helps manage competing responses, but it's not responsible for activating a new script following the punch line. That responsibility is shared across our entire brain, which holds all the knowledge necessary to know what jokes actually mean. So, conflict may be essential for humor, but we won't find a joke funny without some resolution. Indeed, without resolution, we get no pleasure. It's the difference between telling a waitress we're on a diet, and expecting that a knife will miraculously make calories disappear.

## BEYOND THE STAGES

It's important to note that the *constructing, reckoning,* and *resolving* stages aren't just a way of looking at humor. They reflect common beliefs about how we process *all* aspects of our environment. We humans are always guessing and jumping the gun, just as we're always dealing with conflicts and looking for ways to resolve them. Jokes are merely a specialized way of dealing with these stages very rapidly.

This isn't to say that jokes can't involve multiple stages occurring simultaneously. Naturally occurring humor frequently mixes up the three stages, as we sometimes see in humorous newspaper headlines. "Red Tape Holds Up Bridge," claims one. "Doctor Testifies in Horse Suit," claims another. These headlines are each worthy of a Jay Leno stand-up routine, but their most impressive aspect is that in just a few words each calls on us to simultaneously construct, reckon, and resolve conflicting interpretations. It's not enough for a headline to be merely ambiguous, because if this were the case, then "Doctor Testifies in Suit" would be just as funny. Rather, it's that the unintended meaning brings us all the way to imagining a lawyer dressed as a horse and wondering how long it will take before the judge declares contempt of court!

Before going on to the next chapter, let's revisit our old friend the anterior cingulate—the brain region that manages conflict by listening to all the voices in our head and telling the unwanted ones to shut up. We know that in a brain populated by billions of neurons, the anterior cingulate is a conflict mediator, a sort of United Nations surrounded by countries that frequently disagree. Clearly, it didn't evolve just so we can find jokes funny. On the contrary, it multitasks, and nowhere is this more apparent than when we're examining its role in political beliefs.

Colin Firth, the English actor who won an Oscar for his portrayal of King George VI in *The King's Speech,* isn't the kind of guy you'd expect to be conducting a serious academic study. And, for that matter, politics isn't a topic you'd expect to be researched at the University College of London Institute of Neuroscience. This makes the study Firth conducted with Geraint Rees, the institute's director, doubly surprising. The idea for the study came to Firth when he was asked to participate in a guest editorship program at the BBC. Firth asked Rees to scan the brains of British conservative Alan Duncan and Labor Party leader Stephen Pound, because he wanted to see if it was possible to differentiate their brains based on their opposing political beliefs. "I took this on as a fairly frivolous exercise initially," says Firth. "I mean, I just decided to find out what was biologically wrong with people who don't agree with me—and see what scientists had to say about it."

Various parts of the politicians' brains did indeed light up when they talked about their jobs. This in itself made for some entertaining anecdotes to share on the air, but even more interesting was what happened when Firth and Rees extended the experiment to a wider sample of ninety randomly chosen subjects. Specifically, they asked subjects to identify their political orientation along a 5-point scale, from very liberal to very conservative, and then put them in a scanner and measured the size of two structures within their brains: the amygdala and the anterior cingulate.

The first thing Firth saw is that the anterior cingulate in liberals' brains was far bigger than in conservatives'. And conservatives? Their amygdalae were bigger than for liberals. We haven't talked much about the amygdala yet, but it's part of the reward circuit, which delivers dopamine throughout the brain. There's something else it's responsible for too—fear, especially as it's related to learning and making decisions. So, by showing that conservatives have a larger amygdala and that liberals have a larger anterior cingulate, Firth and Rees demonstrated that these individuals are likely specialized for different things. Liberals are more highly tuned for conflict detection. Conservatives, for emotional learning.

This difference was big enough that Rees and Firth were able to correctly classify subjects as either very liberal or very conservative with 72 percent accuracy just by looking at their brains. By contrast, religious intensity—one of the most influential factors in political belief—predicts liberal or conservative leanings at only about 60 percent.

Speaking of religion, you might be surprised to know that this is linked with anterior cingulate activity too. One study at the University of Toronto found that when religious people think about God, activity in their anterior cingulate decreases, suggesting that, for them, spirituality is a conflict-reducing process. The exact opposite result was found among atheists, whose anterior cingulate activity *increased* when they thought about God. For atheists, faith in a supernatural higher power doesn't resolve conflict. It increases it.

》 《

Does this finding mean that liberals and atheists are wired to be funnier people? Probably not. What it does suggest, however, is that liberals are more attuned to noticing conflict. And given that the anterior cingulate helps resolve ambiguity, liberals might also be more capable of adapting to complexities and contradictions. Conservatives, on the other hand, are probably more emotional. They tend to resolve complexity through their feelings—which isn't a bad thing either, because without feelings, humor wouldn't exist.

The anterior cingulate and the amygdala surely evolved for a reason other than just to identify funny jokes. They help us make sense of our world by seeking out conflicts and complexities, and then by resolving those conflicts in an emotionally satisfying way. Liberalism and conservatism, like jokes and religion, are just different ways of dealing with confusion. And without that confusion, we would never laugh.

# 3

## >> Stopover at the Empire State Building

*Man alone suffers so excruciatingly in the world
that he was compelled to invent laughter.*
— FRIEDRICH NIETZSCHE

*"Other than that, how did you like the play, Mrs.
Lincoln?"*
— UNKNOWN

IF SEPTEMBER 11, 2001, WAS THE DAY THAT FOREVER CHANGED American politics, then September 29, 2001, was the day that forever changed humor.

Most people don't think of that day as particularly special, but New Yorkers know better. It wasn't the day that marked the United States' invasion of Afghanistan, which wouldn't happen for another week. And it wasn't passage of the Patriot Act, which was still over a month away. No, September 29, 2001, was the premiere of *Saturday Night Live's* twenty-second season.

Just as we all remember the tragic events of 9/11, we also recall the somber mood that followed. Television stations stopped showing sitcoms and anything other than twenty-four-hour news. Musicians canceled concerts, professional football and baseball games were suspended, and for only the second time in its fifty-six-year history, Disney World closed its doors. As Gilbert Gottfried discovered when he tried to make a joke about the tragedy at the Hugh Hefner roast, the country wasn't yet ready to laugh.

The challenge facing Lorne Michaels, the producer of *Saturday Night Live* with its audience of millions, was enormous. Eighteen days after an incident that took the lives of more than twenty-five hundred New Yorkers, over four hundred of whom were police, firefighters, and paramedics, he was supposed to air a show whose sole purpose was—comedy. Nobody would have blamed him if he had canceled the show. Only a handful of entertainment programs were back on the air, yet Michaels knew that *Saturday Night Live* was special. It represented the city itself, and if the premiere didn't air on time, an unacceptable message would be sent to the rest of the country.

"What am I doing here?" asked the actor Stephen Medwid, an extra for the show who was scheduled to audition for a talent coordinator only two days after the tragedy. Sirens still blared throughout the city, and through a window he could see the smoldering afterglow of ground zero. "The only answer I could come up with was: maybe laughter is the best medicine."

When the show aired, it opened with New York City mayor Rudy Giuliani standing center stage, surrounded by two dozen members of the New York City Fire and Police Departments.

"Good evening. Since September 11th, many people have called New York a city of heroes. Well, *these* are the heroes. The brave men and women of the New York Fire Department, the New York Police Department, the Port Authority Police Department, Fire Commissioner Tom Von Essen, and Police Commissioner Bernard Kerik."

Then, after a brief discussion about the heroism of those who perished, Giuliani introduced Paul Simon, who began singing "The Boxer," a song about New York City originally recorded only a short distance away at St. Paul's Chapel. When the song concluded, the camera returned to Giuliani, who was now standing next to producer Michaels.

"On behalf of everyone here, I just want to thank you all for being here tonight. Especially you, sir," Michaels told the mayor.

"Thank you, Lorne," Giuliani replied. "Thank you very much. Having our city's institutions up and running sends a message that New York City is open for business. *Saturday Night Live* is one of our great New York City institutions, and that's why it's important for you to do your show tonight."

Michaels paused.

"Can we be funny?"

Though Giuliani wasn't a comedian, he certainly knew how to work a camera. His delivery was as deadpan as you get.

"Why start now?"

It wasn't a joke that made people laugh out loud, but it was a line that everyone remembers. We desperately wanted permission to laugh again—and only approval from the mayor of the city could have made that possible. Somehow, Giuliani made it feel as though we were being unpatriotic if we didn't.

I start this chapter with the story of *Saturday Night Live's* return following the terrorist tragedy because it shows how sensitive laughter can be. Still, the show wasn't particularly cutting-edge that night. For example, the opening monologue was supposed to begin with the guest host, Reese Witherspoon, telling a joke about a polar bear cub:

> *There once was a polar bear couple who had a beautiful polar bear baby. He was the cutest baby, and could run really fast and talk very early. His first question to his mother was "Mom, am I a real polar bear?" And his mother says, "Of course you're a polar bear. I'm a polar bear, and your daddy's a polar bear, so of course you're a polar bear."*

*So the baby bear keeps growing, learning how to fish and making his parents very proud. Then, after a few months again he asks, "Mom, are you sure I'm a polar bear?" "Yes, honey, we're polar bears," answers his mom. "Your grandma and grandpa are polar bears. You're pure polar bear." And he says, "Okay."*

*Then, on his first birthday, his parents throw him a huge party, telling him how proud they are of him, and just as he's about to blow out the candles to the cake he asks, "Mom, are you sure that I'm 100 percent pure polar bear?" The mother, flustered, asks, "Why do you keep asking that? Of course you're pure polar bear!"*

*"Because I'm fucking freezing!"*

Up to the moment the show aired, Witherspoon worried about the joke, particularly the ending. Michaels pleaded with her to tell it, profanity and all. He offered to pay whatever fines the FCC lobbied against her, saying that it was worth the cost to prove to the viewers that New York City was back and running. Witherspoon understood the argument, but she still changed the ending. "I'm freezing my balls off!" she said. Everybody laughed, and nobody knew she had censored the joke, though the effect wasn't the same.

Humor is about emotion as much as surprise. When jokes go too far or use offensive language, we feel uncomfortable. That discomfort is why the crowd booed Gottfried's Aristocrats joke, and why Witherspoon chose not to say "fuck" on national television. But sometimes a little discomfort is a good thing. It's useful not just for solving insight problems and getting punch lines but also for turning our stress and negative emotions into something positive, like laughter.

This chapter explores why.

## HUMOR GETS A BAD RAP

Surprisingly, for much of our history humor has been quite unpopular. Plato outlawed humor in the *Republic,* claiming that it distracted people from more serious matters. He wasn't alone; the ancient Greeks,

as enlightened as they were, believed that laughter was dangerous because it leads to a loss of self-control. Thomas Hobbes was a bit more
practical, claiming that humor is a necessary part of life, but only for
people of inferior intellect. It gives them an opportunity to feel better
about themselves, he claimed, especially when pointing out the imper-
fections of others.

Philosophers aren't the only ones antagonistic toward humor. The
Bible is downright aggressive. On several occasions the Old Testament
mentions God laughing, but almost always as a form of derision or
scorn, such as in the Second Psalm:

> *He that sits in the heavens shall laugh*
> *The Lord shall have them in derision*
> *Then shall He speak unto them in His wrath*
> *And vex them in His sore displeasure.*

Not the kind of laughter anyone wants to hear. Throughout the Bible,
when people laugh it's generally out of foolishness, such as when Abra-
ham and Sarah laugh at the idea that they could conceive a son. Some
researchers have gone so far as to count the number of times that God
or his followers laughed, characterizing each instance as due to aggression, sadness, or joy. The winner by a landslide was aggression, at 45
percent. Laughter due to joy occurred only twice.

And for those who argue that the Old Testament is inherently
darker than the newer version, consider this: there are several on-
going debates among religious scholars over whether Jesus laughed.
Not as recorded in the Bible. In his entire life.

Why has humor been treated so harshly throughout history? One
reason is that humor is inherently subversive. Some jokes are innocu-
ous, with topics like chickens crossing roads and elephants hiding in
cherry trees, but most humor isn't like that. It treats serious subjects
with frivolity, and sometimes with rudeness and inconsideration too.
Consider the following joke, which I heard many times during my
childhood but is probably new to the current generation:

*What does NASA stand for? Need another seven astronauts.*

Most people won't get the punch line until I tell them that this joke was popular in 1986, following the explosion of the space shuttle *Challenger.* Seventy-three seconds after taking off from Cape Canaveral, an O-ring in the ship's rocket booster failed, causing a fuel leak and the breakup of the aircraft. All seven passengers perished, including the teacher Christa McAuliffe, who was riding along as part of NASA's "Teachers in Space" project.

This wasn't the only *Challenger* joke, either; there were quite a few. They didn't appear right away but, rather, a couple weeks following the incident. One study identified the latency period between this particular tragedy and the corresponding joke cycle at seventeen days. The death of Princess Diana had a shorter latency period. The World Trade Center disaster had a much longer one.

Our fascination with dark humor is shown by the immense variety of sick jokes, including *Challenger* jokes, AIDS jokes, and Chernobyl jokes—to name just a few. Whole generations of jokes have even outlived the tragedies that spawned them. When I was growing up, everybody had their favorite "no arms or legs" joke. *What do you call a kid with no arms and no legs nailed to the wall? Art. What do you call a kid with no arms and no legs floating in a pool? Bob.*

What many readers may not realize is that there was once an entire generation threatened by this very affliction. Thalidomide, frequently prescribed by doctors in the 1950s and 1960s, had a terrible record of causing a wide range of birth defects. One of those defects was phocomelia, the congenital absence of limbs. Since the primary reason for thalidomide's use at the time was the treatment for morning sickness, thousands of children were affected. The survival rate of phocomelia was about 50 percent, so there probably were babies born without arms and legs, and their names could have been Art or Bob.

Some people claim that these jokes highlight the worst aspect of human behavior. AIDS jokes, they say, are nothing more than an excuse for homophobia. Thalidomide jokes make fun of the handicapped. One reviewer even claimed that *Challenger* jokes encouraged young students to make fun of teachers. But others don't think these assertions are fair. They believe the truth is more complicated than that—and, not surprisingly, their argument has to do with the different ways our brains deal with conflict.

"I'll tell you one thing. [These jokes are] not a form of grieving," says Christie Davies, British humor researcher and author of more than fifty books and articles on the topic. If there's one person who can explain the purpose of sick humor, it would be him. He has given presentations about the topic in over fifteen counties, appeared internationally on television and radio, and even testified before the Supreme Court. In short, when it comes to sick humor, Davies knows what he's talking about. And he's not easily offended.

"The second thing they aren't is callous. The explanation, I believe, is incongruity." Davies's theory, and the one supported by most humor researchers, is that despite the cruel or insulting nature of sick jokes, the teller's intention doesn't have to be vile. In fact, to understand the true message in sick jokes, we have to explore the incongruous feelings behind them. When tragedy strikes, we may have many reactions. We may feel sadness, pity, even despair. We may also feel frustration over the manipulation of our emotions by news reporters, particularly on television. In short, we experience conflicting emotions. Some people argue that sick jokes elicit feelings of superiority, which perhaps also is true, but this contention doesn't explain why coming up with alternatives to the acronym AIDS is funny to some people but yelling "Ha, ha—you're sick!" in an oncology ward is funny to nobody at all. We laugh at jokes about groups or events only when such jokes bring about complex emotional reactions, because without those reactions we'd have no other way to respond.

Some readers may be concerned that viewing sick humor as the result of conflicting emotions is dangerous, because it means that laughing at these jokes isn't cruel but, rather, simply a means of addressing our feelings. It may even seem like an open invitation to tease the sick, dead, or handicapped. But it's not.

The best evidence that sick humor doesn't have to be perceived as offensive comes from a study of jokes that seemingly made fun of the very people it used as subjects. Conducted by the psychologists Herbert Lefcourt and Rod Martin, it involved thirty disabled persons who were asked to view a series of cartoons about people with disabilities. For example, one cartoon showed an elevated gallows. On one side was a set of stairs leading up to the noose, and on the other was a wheelchair ramp next to a handicap placard. Another cartoon showed a cliff with a sign reading "Suicide leap." Next to the ledge was a wheelchair ramp and a handicap sign.

The experimenters didn't want the subjects to know that the purpose of the study was to assess their sense of humor, so they showed the cartoons casually while preparing the room for interviews that were to follow. After surreptitiously noting the subjects' reactions, they administered a series of questionnaires and surveys about their feelings regarding being disabled.

Lefcourt and Martin found that the subjects who laughed most at the jokes were also the ones who were better adjusted to their condition. Compared to other subjects, they exhibited higher levels of vitality, more self-control, and better self-concepts. In short, those who viewed their disabilities in the healthiest manner found the jokes funniest.

These results aren't surprising in light of other research showing that bereaved widows and widowers who are able to laugh about their loss are observed to be happier, better equipped to deal with stress, and more socially adapted. Women who use humor as a coping mechanism after undergoing surgery for breast cancer also demonstrate reduced postsurgical distress.

Further evidence that sick jokes don't have to be offensive in order to be funny comes from manipulation of sick jokes themselves. These jokes vary in a number of ways, not just in their targets but also in their degree of cruelty and fit—and by manipulating each of these factors, researchers can determine whether using vulnerable targets makes jokes too offensive to be appreciated (e.g., *How do you prevent a dead baby from exploding in a microwave? Poke holes in it with a coat hanger*).

Just such experiments have been conducted—for example, by Thomas Herzog of Grand Valley State University in Michigan—and from them we see two interesting things. First, cruelty doesn't improve funniness. Jokes perceived as most vile (e.g., those involving dead babies) are usually seen as least funny. But so are jokes rated especially low for cruelty, many of which aren't emotionally engaging at all. So, cruelty doesn't make the jokes funnier, it only provides a means for introducing emotional conflict. Too little edginess—too little emotional conflict over whether a joke is appropriate or not—and the punch line fails. Too much edginess, and there's no conflict because the inappropriateness is clear from the start.

Second, we see that the biggest predictor of humorousness is fit. That's defined by how well the punch line leads to both incongruity and resolution (much like the *resolving* stage discussed in Chapter 2). In other words, the more effectively the punch line leads to a surprising ending, the funnier it is. It's not enough that we be shocked or surprised. Our humor must bring us someplace new, emotionally as well as cognitively.

Part of the reason why there are so many kinds of sick jokes is that our minds are confronted with mixed emotions in so many ways. For example, we feel sorry for people with handicaps, but we also want to empower them and treat them as they should be treated—like everyone else. And though we grieve the victims of natural disasters, we may simultaneously feel manipulated by the media for telling us how to feel. Television, in particular, can be both an overwhelming source of information and a major source of conflict, primarily because it's so immediate. What do we do when a disaster strikes? We turn on the TV.

"The television will try to convince you on the spot of the emotional impact of the situation, and the rhetoric is all about the immediate," says Davies. "But you can't feel through the television what people are feeling at the scene right away. What you're seeing on the screen is sanitized, but the guy describing it is saying how awful it is. You're looking on the screen and part of you is recognizing that this is absurd."

It turns out that immediacy is a big issue when it comes to humor. As noted earlier, it took seventeen days for *Challenger* jokes to hit campuses and playgrounds. That comes to about two and a half days of grieving per lost person. By that calculation, more than seven thousand days—or nineteen years—should pass before anybody laughs about the attacks on the Pentagon and World Trade Center. Though I suspect we can't trust such a simple formula, it might not be so far off. The movie *Flight 93*—which wasn't even a comedy but a dramatic re-creation— didn't hit theaters until almost five years after the tragedy. And though the satirical periodical *The Onion* did joke recently that Americans should honor 9/11 by not masturbating on its anniversary, such humor is rare and typically respectful of the victims.

Still, many jokes did appear right after 9/11—not within mainstream media but on the Internet. These were also some of the most jingoistic and violent. Consider, for example, the Photoshopped picture of the Statue of Liberty holding a decapitated head of Osama bin Laden. Or the picture of a 747 being flown into the heart of Mecca with the caption "Don't get mad—get even." It's hard to mistake the emotional messages these pictures were meant to convey.

But the really important aspect of these 9/11 jokes is that they reveal our true feelings about the incident. There's anger, of course, but also frustration and occasionally irreverence. One cartoon that comes to mind portrays several Teletubbies jumping from the burning Twin Towers, with the caption "Oh no!" Another depicts a mouse cursor hovering over the World Trade Center, next to a computer message window asking "Are you sure you want to delete both towers?" These jokes didn't make fun of terrorists. They made fun of the grieving process itself. And, as noted, they appeared right away, while television stations were still canceling award shows and Lorne Michaels was struggling with how to host a live comedy show. They weren't sympathetic or sentimental; they were the opposite of that.

In short, they reflected what people wanted to say: "Don't tell me how I'm supposed to feel. I can recognize a tragedy when I see one without being reminded about it by around-the-clock news coverage."

Such jokes reveal something new and remarkable about the human mind—namely, that being told we can't laugh makes us *want to laugh*. It makes us want to Photoshop a picture of a huge gorilla grabbing planes from the sky near the Twin Towers with a caption reading "Where was King Kong when we needed him?"

The human brain is an obstinate beast. It doesn't like being told what to do.

## SCARY MOVIES AND RELIEF

When you think of emotional complexity, you probably don't think of movies like *The Exorcist* or *Salem's Lot*. These are horror flicks, and they're intended to convey a specific feeling—fear. They're quite good at it too, which raises the question: Why do people watch them? If horror movies elicit feelings that most of us try to avoid, why would we pay to see them in theaters?

My wife enjoys horror movies, but I don't. It's not that I'm a fearful person, or at least I hope I'm not. Mostly, it's because I don't see the fun in them. But my wife, like so many others, disagrees. She says there's great fun in being terrified. She loves roller coasters too, which makes no sense to me at all.

I had always assumed that horror movies are popular because people like the relief that follows the scary scenes. This belief stems in part from the relief theory of humor, which states that we laugh when we are suddenly released from threat or discomfort. The idea is rooted in Freud's psychoanalytic theory, which holds that everything we do (including laughter) is affected by stress imposed by our superego on the hedonistic drives of our id. But this theory is unsatisfying for several reasons, one being that it's about as scientific as palm reading and astrology. It also fails to explain why we don't spend our days hitting ourselves over the head with hammers just to enjoy the satisfying feeling of stopping.

Apparently, I'm not the only one who questions the sanity of horror movies. Eduardo Andrade, professor of marketing at the Hass

School of Business at the University of California–Berkeley, is also uneasy with this popular genre—or, rather, with the explanations that
people give for why horror movies are so popular. For instance, why
do we need to watch scary or disgusting scenes to feel good about ourselves? If viewers supposedly enjoy these movies because of the relief
that comes when they're over, why do the bad guys almost always win
in the end and return in the sequel? And if it's true that people who
like horror movies are less sensitive than everybody else, and aren't
bothered by gruesome scenes, why aren't these differences revealed
on personality tests?

To seek an explanation, Andrade performed a series of experiments that only horror-movie fans could love. He showed his subjects
ten-minute clips from two popular horror movies: *The Exorcist* and *Salem's Lot*. In one version of the experiment, he asked subjects to rate
their joy and discomfort, otherwise known as positive and negative
affect, both before and after each film was viewed. In another, he asked
for these assessments during the scariest scenes. Reported feelings were
assessed for those who considered themselves horror-movie fans, compared to those who generally avoid this type of film.

Contrary to popular expectations, both fans and nonfans reported
increases in discomfort after viewing these films. In short, *all* of the
subjects found them disturbing, regardless of whether they liked them
or not. The experimental differences showed up when Andrade looked
at the subjects' feelings of joy. Fans reported increased joy, along with
discomfort, during the scariest scenes. That happiness continued up to
the end of the movie clips, whereas nonfans showed no such enjoyment. As soon as things got scary, they were ready to close their eyes
until the experiment was over.

Andrade's data revealed that horror-movie fans actually experience
two emotions at once—joy and fear.

When I first read about these findings, I went with my wife to a horror movie to conduct a test of my own. I didn't know what I expected
to see, but as we walked into the early showing of *The Innkeepers*, I
promised to keep my eyes open. For the first time in my life, I hoped

that the people in the theater would be loud and disruptive so that I could see their reactions.

During the first scary scene, many people in the audience screamed. Quiet lulls were punctuated by anticipatory gasps, and then the screams came again. At first it seemed as though this would be the extent of the audience's reaction, but then something strange happened. During the next scary scene, one in which the main character enters a possessed cellar, rather than screaming several people laughed. There had been no joke, only a bunch of very startled birds, but these people laughed anyway. This happened several more times, particularly during scenes for which there was an especially long build-up of suspense.

What made the audience members laugh? And why did the subjects in Andrade's study report joy during scenes of terror? One of Andrade's last experimental manipulations provides the answer to both of these questions. Before the horror movies were shown, he presented brief biographies of the principal actors to the subjects to remind them they were observing people "playing a role." He also placed pictures of the actors next to the screen during the movies. This time, even the nonfans reported feelings of happiness during the scary scenes, to almost the same degree as the fans. Evidently the pictures and biographies provided the subjects with what Andrade called a "protective frame" from which to view the events playing out on the screen. Being reminded that they were merely watching a movie allowed them to override their fear so that enjoyment could emerge. It's as if the subjects wanted to enjoy the movies but their fear wouldn't let them. The experiment allowed these latent feelings to finally be free.

Harald Høffding was a Danish philosopher, well known in the late nineteenth and early twentieth centuries but relatively unfamiliar to most people today. He made one immense, though often overlooked, contribution to the study of humor that is worth mentioning now— what he called "Great Humor." Emotions, claimed Høffding, reveal themselves in many forms, typically in the form of "simple" feelings, like happiness or sadness. But sometimes our emotions fuse and form organized complexes reflecting entirely new perspectives. This happens at what he called "the heights of life," moments when we become so in tune with our emotions that we act based on the totality of our experiences rather than in response to the immediate emotional or cognitive demands placed on us by our surroundings.

Høffding's thoughts on Great Humor serve as a useful introduction to the idea of emotional complexity. Indeed, Great Humor reflects an appreciation of life not just from the perspectives of happiness or sadness but more totally, through a complex synthesis of these emotions. The best humor doesn't make us feel merely one way or another. It's more than that. It makes us laugh at the handicap joke at the same time that we feel sympathy for its target.

"It's pointless thinking of humor in terms of catharsis. We're not talking about washing away anything," says Davies, and I agree. Sick humor doesn't simply purge emotional stress from our system—it "cultivates" the stress so that we can achieve some sort of resolution. This gets at the difference between the Greek concepts of catharsis and cathexis. Catharsis is the cleansing of one's feelings. Cathexis is the lesser-known opposite of that. It's the investment of emotional energy, the revving up of the libido. Humor provides us with relief, not by washing away bad feelings but by activating them, along with positive ones so that we can enjoy a complex emotional experience.

Nowhere is this more apparent than in gallows humor, the darkest form of comedy in which one makes light of dire circumstances. When the remaining members of the Monty Python comedy troupe sang "Always Look at the Bright Side of Life" at the funeral of their deceased friend Graham Chapman, they weren't celebrating his death—they were celebrating his life. Indeed, tragic environments, such as funerals and war, are effective breeding grounds for humor because they provide the same release as horror movies, allowing participants to confront their emotions head-on. *Why hasn't Hitler invaded England yet?* asked one Czech in the fall of 1940. *"Because the German officers haven't managed to learn all the irregular English verbs yet."*

Perhaps the most famous example of gallows humor is the story of Gerald Santo Venanzi, an Air Force captain from Trenton, New Jersey. During a bombing raid over North Vietnam in September 1967, Venanzi's RF-4C aircraft was shot down near Hanoi. Almost immediately he was taken prisoner and, along with several other soldiers, was subjected to brutal treatment. Many of his cohorts were in worse shape than him, bound and miserable and desperate for something positive. Seeing the bleakness of the situation, Venanzi did the only thing he could: he created an imaginary motorcycle, as well as a fictitious chimp named Barney.

To amuse his fellow prisoners of war, Venanzi "rode" the bike around the prison complex whenever he could, performing tricks and even taking the occasional spill, as one would expect during such daring maneuvers. Most of the guards thought him mentally unbalanced, but his fellow prisoners loved it. Soon Venanzi began providing sound effects, and eventually the show grew so animated that the guards asked him to stop. It wasn't fair to the other prisoners, they explained, because they didn't have motorcycles too.

Fortunately, after the guards took away his imaginary motorcycle Venanzi still had Barney, who accompanied him to solitary confinement, as well as to numerous interrogations. Barney made insulting remarks about the captors, always directed only to Venanzi but shared with everyone else indirectly. Already convinced that they were dealing with a weak mind, the guards played along, sometimes asking Venanzi to repeat Barney's vile retorts. One guard even offered Barney tea, which Venanzi declined, though he did laugh about this exchange with his fellow prisoners later. Though odd from the perspective of his captors, these antics were the greatest entertainment Venanzi and his fellow prisoners could ever have shared. And the effort eventually earned him the Silver Star, the third-highest decoration a military member can receive during times of war.

Prisoners aren't the only group who routinely use humor to cope with macabre surroundings. Doctors, too, spend much of their days exposed to blood, gore, and general depression—and again their approach to coping seems to be laughter.

What *Catch-22* did for war, *House of God* did for medicine. Written under the pseudonym Samuel Shem, this novel by Stephen Bergman focuses on a group of medical interns struggling to deal with the pressures and complexity of medicine. They create names for uninteresting but acute patients such as "Gomer" (an acronym for "get out of my emergency room") and refer to "turfing" difficult patients to other teams. They even commit suicide and secretly euthanize patients, all in reaction to the extreme stress placed on them by their jobs. Though not as extreme as war, the interns' situation is intense, and lives are saved or lost based on the decisions they make.

Medicine is rife with humor, even where you'd never expect to see it. The textbook *Pathologic Basis of Disease* describes interstitial emphysema as a bloating of the subcutaneous tissue to an "alarming, but usually harmless Michelin-tire-like appearance." The same textbook warns that, since the chance of acquiring hepatitis from eating oysters is 1 in 10,000, doctors should warn oyster lovers never to consume more than 9,999 in one sitting. The *Merck Manual of Diagnosis and Therapy* classifies discharge of flatus, otherwise known as farting, into three categories: the slider, the gun sphincter, and the staccato.

If these instances seem mild, making it difficult for us to decide whether the humor is inappropriate, consider the following true story about a group of medical professionals debating how to treat a baby born with severe neurological defects. The doctors discussed numerous tests, contemplating all the possible information they might gather about the child's condition. It soon became clear that the situation was hopeless, though no one wanted to be the first to give up. Finally, one of the doctors ended the debate: "Look. He's more likely to *be* second base, than play it."

Wow! Fortunately the parents weren't around to hear this. But that's not the point. The doctor wasn't trying to be cruel. He was doing what the American essayist George Saunders calls "rapid-truthing." The term was originally coined in reference to Kurt Vonnegut's use of blunt, unvarnished humor in the book *Slaughterhouse Five* to describe the depravity of war. It means "to express truth without flourish." As Saunders put it, "Humor is what happens when we're told the truth quicker and more directly than we're used to." In other words, likening a struggling baby to a piece of sporting equipment is funny for the same reason it's horrifying—it expresses an idea so horrible that we're unaccustomed to addressing it so directly.

In most instances of sick humor, the target isn't obvious. Medical humor, particularly of the macabre sort, doesn't make fun of the patients—it makes fun of death. There's an old story, also true, about a group of doctors working late one night in the ER, and together they decided to order a pizza. It was after three in the morning and their delivery still hadn't come, when suddenly a nurse interrupted their complaints by bringing in a gunshot patient. The doctors quickly recognized the patient as the pizza delivery person, who apparently had been shot delivering their order just outside the building. They worked for hours trying to save him, going so far as to open his rib cage and massage his torn heart. But their efforts were futile and the boy died.

Tired and depressed from having lost their patient, one of the doctors finally asked the question that was on all their minds.

"What do you think happened to the pizza?"

Another doctor looked outside and spotted the box, face up, only a few steps from the ER doors. He retrieved the food, then set it on the table in front of his colleagues.

"How much do you think we should tip him?"

What does the question about tipping really convey? I believe it says several things. First, that we're all going to die, and that trivial matters like tipping will remain long after we're gone. Second, that being alive is a pretty special condition, one that shouldn't be wasted, just like pizza. And, third, that death may come and take even the most innocent people, but it can't make us afraid if we don't let it. Death is the enemy, not pizza, and the only way the brain can express all these complex ideas is to laugh.

## JOKES WITH A TARGET

One major benefit of information sources like the Internet is that even nonscientists have access to the latest discoveries in brain science. Consider, for example, mirror neurons—brain cells that fire when we take an action ourselves, but also when we see someone else demonstrate the same behavior. They may fire when someone reaches for food, shakes someone's hand, or picks up a book, but they don't care if we're doing these things ourselves or watching someone else do them at a distance. Mirror neurons were discovered in the 1990s, and they've become familiar to the general public—in part, because they're so amazing. Many scientists now claim that these cells are responsible for recognizing intention in others, and maybe even empathy.

But there's an even more exciting class of neurons, in a neuroscientific sense at least, that are recently getting public attention, and they're known as spindle cells. These cells (scientifically called Von Economo neurons, after the Romanian neuroscientist Constantin Von Economo) are relatively rare, occurring in just a few regions within the brain. One of these is the anterior cingulate. Their appearance is unusual too, four times larger than most other neurons with extremely long projections. And they're found in only a few species other than humans—namely, in our most intelligent ape neighbors, such as gorillas and orangutans, and in a few other advanced mammals like whales and elephants.

What do they do? Whereas mirror neurons are responsible for empathy, spindle cells are responsible for social awareness and emotional control. Research suggests that they perform fast and intuitive updating of feelings and emotional reactions. Their lengthy projections allow them to communicate efficiently across wide regions of the brain, and their emergence after birth—unlike most other neurons, which develop prenatally—suggests that their development is influenced by environmental factors such as quality of social interaction. And, finally, spindle cells are observed only in animals with brains possessing a balance of cognitive and emotional thinking. There's one region each of these animals have in common, the same place in which these spindle cells are most commonly found—the anterior cingulate.

In this chapter we've seen that conflict can be emotional as well as cognitive. When we experience conflicted feelings, we need to reconcile such feelings and establish emotional control. Spindle cells, because they're built for quick, long-range communication, are perfectly suited for this goal.

One thing I didn't mention earlier about the anterior cingulate is that it isn't a single entity. It has parts, much like the larger brain, and one of its most important divisions is between its dorsal and ventral sections (from the Latin words *dorsum* and *ventralis*, which mean "top" and "bottom"). These sections divide the anterior cingulate's cognitive and emotional responsibilities, respectively. The top—or dorsal—part of the anterior cingulate deals largely with cognitive conflict. The bottom or ventral part is emotionally focused.

Returning now to the Stroop task from Chapter 2—reading words in different-colored fonts is a cognitive task, so it typically activates the dorsal anterior cingulate. But there's an emotional version of the task too, called the Emotional Stroop task. For example, rather than asking subjects to report the color of neutral words like *B-L-U-E,* the Emotional Stroop task uses words like *M-U-R-D-E-R* and *R-A-P-E.* The shocking words are unrelated to the task, but the ventral part of the anterior cingulate notices them anyway. Then it sends a warning to the rest of the brain—do not trust this researcher!

It's likely that spindle cells are key for sharing such warnings. They appear to be concentrated in the ventral part of the anterior cingulate, which is responsible for detecting such emotional messages, and they're also found in a region called the fronto-insula, another key area for processing emotion. This makes them well suited for dealing with conflict in social situations, particularly those involving mixed or contradictory emotions. Both the ventral part of the anterior cingulate and the fronto-insula are also active during times of empathy, guilt, deception—and humor. In short, these two brain regions are especially involved in dealing with messy feelings.

This issue of messy feelings is significant because it brings us to another kind of conflict important for humor—personal conflict. Sometimes the target of our jokes is unnamed, but much of the time our comedy is directed at particular individuals. In these cases the humor is personal, involving feelings about specific people, and maybe even insults. Though we haven't yet identified precisely which neural responses are involved in these kinds of interactions, with the discovery of spindle cells we may not be far off. Their long projections and their connections to our emotional centers allow these cells to access a wide range of feelings, thus helping us to work through our complex emotional reactions. One way to do that is to tell a joke.

*One day, the secretary to Israeli Foreign Minister David Levy hears on the radio that a lunatic is driving against traffic on the busy Jerusalem/Tel Aviv highway. Knowing that this is her boss's route, she immediately calls his car to warn him. "Just one lunatic?" he screams back at her. "They're all driving against the traffic!"*

This joke depends on your knowing who David Levy is, which many readers will not, but I include it here because it highlights two kinds of humor. First, it's clearly an insult. At the end of the joke, we get some clear impressions of Levy—that he's a bad driver, not particularly smart, and maybe hard-headed too. The joke is also political satire, since Levy is a public figure. And it's a polarizing one, which makes the joke even funnier.

During the late twentieth century, David Levy was a prominent and controversial Israeli politician. With only an eighth-grade education, Levy began work in construction before later aligning with his country's moderate right-wing Likud Party. After holding several ministerial positions he attained a well-recognized position in government, but several unfortunate personal characteristics kept holding him back. One was that he frequently appeared stern-faced and pompous. Another is that he never learned English, making international relations difficult. Even in his native language he often made slips of the tongue that caused him to look dense. Eventually he became a symbol of stupid, selfish politicians who are willing to say whatever the public wants to hear.

Enter the wave of *Bedichot David Levi,* which is Hebrew for "David Levy Jokes." This phenomenon, as recorded by Hagar Salamon of the Hebrew University of Jerusalem, took Israel and neighboring countries by storm. They made fun of Levy's intelligence, his arrogance, and— most of all—his inability to recognize his own shortcomings. The jokes became so widespread that a *Los Angeles Times* reporter even wrote an article wondering if he could overcome them. Here's an especially popular one: *A man approaches David Levy and says, "Have you heard the latest David Levy joke?" "Excuse me," David Levy responds. "I'm David Levy." "That's okay," the man replies. "I'll tell it slowly."*

At first glance, this may appear to be just another wave of political humor directed at an easy target. In the early 1990s, nearly everyone was telling Dan Quayle jokes for seemingly the same reasons. The names Quayle, Clinton, and Palin could easily be used in Levy's place for any of the jokes and be just as funny. Or could they?

Though language differences make it hard to answer this question, a closer examination shows that the humor behind these jokes is more complicated. For one thing, Levy wasn't born in Israel but in Morocco— and as a Moroccan Jew, he represented a new faction in Israeli politics. Up to Levy's time, Israel had been dominated by European Zionism; however, as Levy came to power, Jews from eastern, traditionally Muslim countries were beginning to alter the ethnic balance of Israeli society. The fact that Levy frequently emphasized his ethnic origins only heightened the growing tensions associated with this change. Aspects of Levy's personality clearly made him easier to ridicule, but the conflict that people felt about their old and new society played a large part too. Levy frequently complained that jokes about him were motivated by latent racism, and perhaps the accusation worked because the joke cycle eventually ended.

〉〉 〈〈

It's easy to find similar jokes targeting popular American figures, though most of these have nothing to do with racism. By the late 1980s and early '90s, when Dan Quayle jokes were at the height of their popularity, America had become obsessed with money and power. The Reagan years had ushered in an era when wealth was the ultimate status symbol, whether earned or inherited, and Quayle was the perfect example of the latter, coming from two generations of wealthy publishers. Though moderately successful on his own, he was seen as out of touch, dense, and disconnected from common America. It certainly didn't help that Quayle didn't know how to spell *potato,* but at a time when America was ready to rebel against the rich and privileged, he had been set up to fail.

Waves of Clinton jokes in the 1990s. Sarah Palin jokes in the 2000s. Each of these targets tapped into an aspect of society characterized by conflict, whether it was Clinton's chronic infidelity during a time of economic prosperity or Palin's intellectual deficits contrasted with her folksy populism. Why did Americans revel in these joke cycles but leave another figure, Jimmy Carter, relatively alone? Carter didn't escape unscathed, of course—several jokes about peanut farmers surfaced in Washington in the late 1970s. But considering his failures, such jokes were relatively few. Rising gas prices, inflation, and the Iranian hostage crisis led critics to assess Carter's presidency as one of the most ineffectual in recent history. Yet jokes at his expense were scarce, primarily because Carter was not a man to elicit conflicting emotions. Likable, ethical, and smart, he came across as a generally good guy, better suited for peace missions than for leading the free world.

Political jokes are popular because they feed on the mixed feelings that people have about public figures, but what if those feelings are about larger groups? When politicians such as Barack Obama and Newt Gingrich run for office, they expect to be subjected to ridicule as part of the process, but what about jokes about Mexicans, or Polish people? What do jokes about larger social and ethnic groups say about society?

When I was growing up, Polish jokes were huge. Though grossly inappropriate, nearly every child or adult knew at least one. *What do they print on the bottom of Polish Coke bottles? Open the other end. How do you break a Pole's finger? Punch him in the nose.*

Every country has one or more popular targets. Russians make fun of Ukrainians. Australians make fun of Tasmanians. Canadians make fun of Newfoundlanders. Most of the time the jokes are about being stupid, but sometimes they're about being dirty or uncivilized too. These targets may seem to be chosen haphazardly, or aimed only at low-status groups that threaten the home country's prosperity. Not so.

The amazing thing about stupidity jokes is that they can be found everywhere, though the targets aren't usually groups most disliked within the culture. Instead they're the ones just barely outside the mainstream, the ones on the periphery. We make fun of these groups because they're only slightly different from ourselves, and such humor helps relieve the stress and anxiety associated with living in a pluralistic society. To what extent, really, were Americans threatened by the Polish when jokes targeting this group were most popular? Those jokes came almost a hundred years too late for that possibility to be taken seriously. Rather, people laughed at Polish jokes because Polish people were different from those around them, but not so different to be an actual danger to existing cultural norms.

If this interpretation is wrong and racial jokes really are a matter of picking on the downtrodden, then the content of those jokes shouldn't matter. But again, this isn't the case. Though in America we're equally comfortable telling jokes about Poles, Irish, or Italians being stupid or dirty, overseas things are very different. It won't take much time in an English pub before you hear someone call an Irishman stupid, but you'll be walking a long way along Haverstock Hill before you hear a Brit calling an Irishman unclean. Grooming habits just don't show up on the list of things that Londoners tease foreigners about. So what gives?

The clearest example of insult jokes saying more about the teller than the target is the prominence of "dirty jokes" in America, but almost nowhere overseas. *Why do Italian men wear mustaches? To look like their mothers.* You'd never hear that joke in Switzerland, even though the Swiss frequently make fun of Italians. Why the difference? "The key is that Americans and Canadians are obsessed with cleanliness. It's a central value," says Christie Davies. "In Britain and elsewhere, it's something empirical. It matters more how clean you are depending on the circumstances. It's utilitarian. Things ought to be clean because the consequences of being unclean are bad. In America it's a moral value."

Nobody in Britain makes fun of the grooming habits of Irish or Belgian people because nobody there cares. This is one more way that insult humor says more about the tellers of jokes than about their targets, because it shows what their values truly are. It's not that Italians or the French are pathologically dirty and don't own razors, it's that Americans are obsessively clean. To us, everybody is dirty, making such jokes less about insulting others than about coping with our own feelings about personal hygiene.

Arguments like these might sound like unscientific conjecture, and in a way they are, but they're still important for anthropologists and sociologists because humor trends are difficult to quantify. Analysis can go overboard, too. Consider the review published by Roger Abrahams and Alan Dundes in the 1960s. Abrahams was an English professor, Dundes a folklorist, and together they examined a growing wave of humor making its way across America: elephant jokes. "One cannot help but notice in this regard that the rise of the elephant joke occurred simultaneously with the rise of the Negro in the civil rights movement," they wrote. "The two disparate cultural phenomena appear to be intimately related, and, in fact, one might say that the elephant is a reflection of the American Negro as the white man sees him and that the political and social assertion by the Negro has caused certain primal fears to be reactivated."

In case you missed it, these authors are implying that white people like elephant jokes because they're afraid of black people. The article goes on to consider several elephantine characteristics supposedly shared by these groups, including phallic ones.

Obviously, some judgment is required here, as well as patience with 1960s' racial terms and stereotypes. Still, just because such analysis can go overboard doesn't mean insult humor doesn't say more about its tellers than its targets. And to see how, let's finish this chapter with an example involving a slightly less sensitive subject. Let's laugh at some lawyers.

*How do you stop a lawyer from drowning? Shoot him before he hits the water.*

*How many lawyers does it take to roof a house? Depends on how thin you slice them.*

*How many lawyers does it take to stop a moving bus? Not enough.*

Have you heard these jokes before? If not these specific ones, you've probably heard others like them. That's because in the last several decades, lawyer jokes have become one of the most popular kinds in this country. A study of humor types in the 1950s found that of the 13,000 jokes found in common circulation, so few were about lawyers that they didn't warrant counting, but by the late 1990s more than 3,000 Internet sites had been constructed for the sole purpose of sharing lawyer jokes, as compared to 227 for doctor jokes and 39 for accountant jokes. That's a big increase.

When lawyer jokes exploded three decades ago, they were notorious for being both popular and amazingly violent. Looking at the three jokes shown above, we see that all involve murder. None have anything to do with the law, and none give hints for why lawyers should be so reviled. Could it be that lawyers are by nature an especially unlikeable group? Maybe, but as numerous sociologists have pointed out, lawyers have a much worse reputation in other countries. The Netherlands, for example, are well known for their antagonistic attitude toward lawyers. Yet among the 34,000 jokes and humorous anecdotes compiled by Dutch sociologist Theo Meder, only 5 targeted this group.

Something changed within the American court system in the 1980s. That decade brought a pernicious wave of litigiousness to the United

States as well as a near doubling of the number of lawyers. Then, on February 27, 1992, Stella Liebeck ordered a cup of coffee at a McDonald's drive-through in Albuquerque, New Mexico. After leaving with her order, she spilled the coffee in her lap, suffering third-degree burns over 6 percent of her skin. Though Liebeck initially sought to settle only for the costs of her medical bills, about $20,000, the case eventually went to court and Liebeck was awarded almost $3 million in compensatory and punitive damages.

Liebeck and her lawyers were merely trying to change the fast-food chain's practices to make their product safe—over seven hundred similar incidents were already on record—but the public was outraged. McDonald's coffee is supposed to be hot. What is the world coming to if spilling fast food on your lap makes you a millionaire? Only in America, where everyone wants to become a successful, highly paid professional, could the public have such contradictory feelings about lawyers. On the one hand, we want to praise them for protecting the innocent, preserving the law, and making a successful living. On the other, we want to feel certain that someone tripping on the sidewalk outside our house can't sue us into oblivion. By taking such a prominent role in society, lawyers have exposed themselves to both admiration and fear. It's a catch-22—love them or hate them, lawyers are here to stay. Our only option is to laugh at them.

In this way we see that humor serves an important social function, helping us deal with grief and resolve conflicting opinions about prominent figures. It may also be a consequence of living in a social society, allowing us to work through our differences in more mature ways than our ancestors did, for example, with clubs and sticks. The Russian psychologist V. I. Zelvys recounts the story of the Dyak tribes of Borneo, who used to engage in frequent battles among themselves—including headhunting. Whenever these tribes went to war, they always began their skirmishes by approaching one another and swearing in the most obscene of ways. The insults were gruesome, replete with promises to remove limbs and shove them up very private places. They were also quite personal, involving offensive remarks about sexual prowess. Similar traditions have been found in ancient North America and Italy, where the ritualized insults even took on a special rhythmic meter, making them a form of poetry. Only after these insult matches concluded were any actual battles allowed to begin.

I wonder whether, sometimes, the insult matches became so lively that the two sides forgot to actually strike each other physically. Like sick humor and jokes about tipping dead pizza delivery boys, these shouting matches served a social purpose, and for the Dyak tribes of Borneo that purpose was to delay violence, at least for a short while. In modern society this purpose has evolved, helping us to deal with the anger and grief associated with tragedy, as well as to integrate conflicting opinions about prominent individuals. It's easy to see the value of humor in these difficult situations, just as it's easy to see how doctors must sometimes joke about their most helpless patients. Humor doesn't have to be cruel, and it doesn't have to be hurtful either. Sometimes it's simply the only available way to react.

# PART TWO

# "What For?"

## HUMOR AND
## WHO WE ARE

# 4

## >> SPECIALIZATION IS FOR INSECTS

*A person without a sense of humor is like a wagon without springs. It's jolted by every pebble on the road.*

—HENRY WARD BEECHER

*I have a fine sense of the ridiculous, but no sense of humor.*

—EDWARD ALBEE

IT'S TIME TO SHIFT GEARS. IN THE FIRST THREE CHAPTERS WE focused on the *What is?* question of humor. What is humor, and why do some things make us laugh and others not? So far we've seen that humor has distinct components, such as conflict and resolution, but now it's time to shift to what I call the *What for?* question: What purpose does humor serve, and why don't our brains adopt simpler means for turning conflict into pleasure? All this arguing within ourselves seems an inefficient way to do business. Surely, if our brains were simpler and worked more like computers, we'd be happier, more jovial people? Not so. And to show why, let's get to know A.K., a sixteen-year-old girl who went to the UCLA Medical Center hoping to find a treatment for epilepsy and left knowing the exact part of her brain that makes her laugh.

## A.K.

"The horse is funny," exclaimed patient A.K. in response to the doctor's question. The doctor had just inquired why she was laughing, and lacking other explanations, this was her best answer. In front of her the doctor held a picture of a horse, and although it wasn't particularly special, it seemed hilarious to her. She didn't know why.

The doctor continued to show her pictures, and to ask her to read paragraphs and move her fingers and arms. While he did these things, another doctor worked just out of sight, nearer her head. A.K., whom we know only by her initials, understood that the second doctor was probing her brain. The doctors were trying to find out why she was periodically experiencing seizures, and the only way to do that was to identify which area of her brain was malfunctioning. What she didn't understand was why her body kept abandoning her control.

A.K. started laughing again. Again, the doctor asked why.

"You guys are just so funny, standing around."

Assistants took note of each interaction as the test continued. Meanwhile, A.K.'s body continued to make unexpected motions. One moment her right leg tingled, the next her arms twitched. Several times she found herself unable to speak and answer the doctor's questions, and she had no idea why. Then, of course, there was the laughing.

The area of A.K.'s brain being probed was the supplementary motor cortex, and the doctors' goal was to map its functions down to the millimeter. They found that the seizures were originating in the medial frontal region of her brain, not far from the part controlling speech and motor movements—as well as laughter. Although A.K. had never experienced laughter during her seizures, her brain apparently had lumped these behaviors together anatomically, with the motor areas affected by the seizures located right next to the areas controlling her laughter.

In this respect, A.K. was lucky; many patients do laugh during seizures, and the results are never funny. One stroke patient from India suddenly found herself laughing, then fifteen minutes later lost sensation in the right side of her body and the ability to speak. Another patient, a forty-seven-year-old man, began laughing following surgery to repair a ruptured aneurysm and didn't stop for twenty years. And then there's Jenna, a twenty-four-year-old patient from Great Britain who has laughed uncontrollably all her life, up to fifteen times a day. "The laughter is like an explosion," she says, describing the condition that has only recently come under control using medication. "One minute all is normal and then I'm just laughing. It's totally natural, but overwhelming too. Very erratic."

The names for pathological laughter are almost as varied as the condition itself, all sounding like evil Harry Potter spells: *enuresis risosa* (Latin for "giggle incontinence"), *fou rire prodromique* (French for "sudden mad laughter"), *risus sardonicus* (Latin again, this time for "devil's smile"). But the most common name is *gelastic epilepsy,* a term adapted from the Greek words for "laughter" and "seizure," which neurologists agree is the most accurate way of describing the event. That's what a seizure is—uncontrolled and excessive activity in the brain. For patients like Jenna, that activity manifests as laughter.

Pathological laughter tells us a lot about the brain because it shows how humor involves the interaction of many different parts. As discussed earlier, laughter is related to humor the same way a symptom is related to underlying disease: it's an external manifestation of inner conflict. Although that conflict often comes in the form of jokes, it doesn't have to. It can be caused by stress, anxiety, or, in cases of pathological laughter, excessive brain activity due to neural insult. The immense number of different modules in the brain, and the many connections between them, allows us to be highly adaptable as a species, but we're also more susceptible to unusual behaviors, such as uncontrollable laughter, because there are so many ways for the brain to break down.

This might be why the symptoms associated with pathological laughter are so varied. For Jenna, the outbursts don't include any feelings of mirth, only laughter. Other patients feel euphoria during spontaneous laughter episodes, and still others confuse pleasure and pain, an unfortunate condition called pain asymblosia. In some cases, pathological laughter is accompanied by cognitive deficits such as reduced intelligence or memory. In others, there are no additional impacts. There are seemingly a hundred different ways to trick our brains into laughing inappropriately, but no way to guess what will happen from one person to the next.

Possessing such a modular and interconnected brain has both advantages and disadvantages. Species tend to be divided into two groups: specialists and generalists. Specialists thrive only in the environments for which they are highly adapted. A good example of a specialist is the Costa Rican katydid, which avoids predators by camouflaging itself as a leaf from the local flora. Take a Costa Rican katydid from its home and pretty soon you have a dead katydid. Specialists aren't limited to insects and other simple organisms; the koala, for instance, is a specialist too. Its diet consists solely of eucalyptus leaves, and so it's found only in eastern and southern Australia. Unless you visit a zoo, you won't find a koala in Europe, or even a short distance away in Tasmania, because the eucalyptus tree is a specialist too.

The fact that humans live in Europe, Tasmania, and even Antarctica shows how we are extreme generalists. Our specialty is intelligence, developed over generations as an ability to adapt to our surroundings. Our brains are the tools of our survival, allowing us to outsmart our environments instead of merely conforming to them. But they also do strange things when broken, such as causing us to laugh inappropriately—an unfortunate consequence of possessing so many working parts.

To liken these survival strategies to cutlery—which I'm sure is the analogy you were thinking of too—if a koala is a bread knife, then a human is the Swiss Army Fieldmaster. The Fieldmaster not only cuts bread, it opens bottles, pops corks, and saws twigs. It's not perfectly suited for any of those tasks, but it does the job, whatever it's asked to do.

Which also explains our extreme variability. With so many features, and so many parts interacting to give us this adaptive intelligence, there's a lot of room for people to differ. To continue with the cutlery example, an unscientific review of the Williams and Sonoma website shows that they offer twenty-four kinds of bread knives. That may seem like a lot, but you can find that many varieties of Swiss Army Knives just within their "flash collection." Those are knives containing USB drives inside. Their total number of options is well in the hundreds.

I'm not trying to sell you a Swiss Army Knife, but it's worth recognizing that complexity has a price—namely, unpredictability. It means that we have more parts to break, and even when our brains work perfectly, they do so in idiosyncratic ways. This chapter explores one consequence of that unpredictability—individual differences. Nowhere are those differences more apparent than in our sense of humor, which is why it remains one of the best ways for examining who we really are.

## STATES AND TRAITS

Humor is notably short on quantitative formulas. Sure, surprise and internal conflict are important, but how can we possibly measure these things? It's impossible. Still, this hasn't stopped some people from trying—for example, Peter Derks, who came up with this quite clever formula:

Humor = salience (trait + state) × incongruity + resolution

At first this looks like a jumbled mess of words. Salience? What's that? But further examination suggests that Derks is actually onto something.

Let's start with the second half of the formula: incongruity and resolution. As we've already discussed, we laugh at things that surprise us (incongruity) and that force us to look at things differently (resolution). These concepts parallel the stages of *reckoning* and *resolving* described earlier. Granted, I would have given resolution a more important role, perhaps making everything else depend on it exponentially, but regardless, from incongruity and resolution we see that we find things funny that catch us off-guard and change our view of the world.

Now let's consider the first half of the formula, which says that humor also depends on salience. Salience has two components: trait and state. When we understand these elements, we should see how all the humor ingredients come together. Right?

I'm 5'9", roughly 180 pounds, and whenever I eat mints I always sneeze at least three times. These details have accurately described me since turning an adult, thus making them traits. They don't change, at least not quickly, and so for practical purposes they can be described as fixed. Contrast these traits with the fact that right now, as I write this paragraph, I'm experiencing a dull pain in my left ankle. This morning, as I let my three dogs outside, my five-year-old rescue mutt Maynard tripped me by running between my legs. As a result, I'm in a bit of a surly mood and trying hard not to blame Maynard for his enthusiasm. That's a state, and it will surely change soon, either when my ankle stops hurting or when Maynard does something funny like roll on his back and purr, as he sometimes does when he forgets he's a dog.

This is a roundabout way of saying that our moods vary from moment to moment, but we possess general dispositions too, and both have a big impact on humor. For example, many religious people have a poor sense of humor. This may seem an unfair generalization, but at least it's a scientifically based one. I'm referring to a study by the Belgian psychologist Vassilis Saroglou from the Université Catholique de Louvain. He gave nearly four hundred subjects a variety of sense-of-humor tests, having previously assessed their self-reported religiosity. He found that strength of religious beliefs was inversely proportionate to social humor, and also that religious men tend to tell self-defeating jokes, perhaps due to their discomfort over the lighthearted nature of humor, given their spiritual convictions.

As you might expect, countless studies have looked at the relationship between personality characteristics and sense of humor. I won't go into most of them because they aren't very informative—how surprising is it, really, that cheerful people tell more jokes than sad people? However, one experiment in particular says something important about how we think, beyond the positive influence of being in a good mood. I'm referring to a study conducted by Paul Pearson, who is both a psychologist and a member of the Cartoonist Club of Great Britain. He administered a personality test—specifically, the Eysenck Personality Questionnaire (EPQ)—to sixty professional cartoonists and found that sometimes the least likely people to tell a joke are the funniest.

The EPQ is perhaps the most widely used psychological assessment there is, so it's worth addressing here. Developed by the psychologist Hans Jürgen Eysenck, with help from his wife Sybil, the test was designed to measure three key aspects of our temperament believed to be set at birth. Though no one, including the Eysencks, believe these three characteristics never change, it's widely accepted that they remain relatively stable over our lifetimes. As such, they're very useful to measure.

The first is extraversion. It exists on a continuum, ranging from introversion to extraversion, and it describes how much energy we seek from our environment, as contrasted with how much we like to be alone. It's also closely linked with arousal—extraverts tend to seek arousal from their surroundings as a means of overcoming boredom, while introverts seek quieter environments due to their jittery natures. If you feel a constant need to be around people and to experience stimulating environments, you're probably an extravert. If that sounds like a lot of work, consider yourself on the other end of the spectrum.

The second is neuroticism, which exists on a continuum ranging from stability to neuroticism. This measures how much anxiety we typically feel and how influenced we are by depression, tension, and feelings of guilt. Neuroticism is closely linked to the fight-or-flight response, a response that is activated relatively quickly in neurotic people, because they're so easily stressed or made anxious. By contrast, people with higher levels of stability tend to be cool under pressure.

The final characteristic is psychoticism, which contrasts with socialization. Psychotic individuals are assertive, manipulative, and dogmatic. This makes them inflexible with their environment, and pushy about it too. They can also be tough-minded, along with reckless and hostile. Testosterone is often identified as the culprit for this behavior, which might explain why studies conducted in more than thirty countries have found that men, on average, exhibit higher levels of psychoticism than women.

It's important to note that these characteristics do not imply any sort of pathology. Psychoticism can be a diagnosis, or it can simply describe someone's position on a much broader continuum. Which is good, because the cartoonists in Eysenck's study scored much higher than the normal population on both neuroticism and psychoticism. So, apparently these artists were a bit "on edge" with respect to anxiety and aggression. What's more surprising is that the cartoonists didn't differ from the rest of the population in terms of extraversion—a surprising finding because if there's one personality characteristic we'd expect to see linked with humor, it's how outgoing we are. Numerous studies have shown that outgoing people tell more jokes, and also appreciate good jokes more. So, what gives—are cartoonists just special?

Apparently not. It turns out that creative people in general show similar results. Professional musicians typically score higher than amateurs on both neuroticism and psychoticism. Painters and sculptors do too, especially on psychoticism, with the most successful artists often scoring highest on this measure. A huge meta-study, examining creative professionals ranging from professional dancers to Nigerian veterinary surgeons, found that one factor best characterized successful scientists and artists alike. That factor was a high degree of psychoticism.

One problem with generalizing on the basis of scientific studies like these is that they often measure different things. Some researchers study professional artists; others, students; and still others, mixtures of both. Some scientists give their subjects questionnaires to assess personality traits, others measure things like laughter. As a result, it's difficult making comparisons without performing years of in-depth analytical work. Thankfully there are scientists like Willibald Ruch willing to do that for us.

Often, researchers conduct studies involving only one or two experiments, with only a couple of different measures, because science takes effort. There's the issue of getting participants, plus practical matters like what tests to administer and how long to gather data before finally publishing. Which makes the following study performed by German psychologist and former president of the International Society for Humor Studies Willibald Ruch all the more impressive. He knew that to understand the complicated connection between personality and humor, he had to be comprehensive. So he didn't just target humor in a single population, he studied more than one hundred adults whose ages ranged from seventeen to eighty-three. He didn't just give one or two personality tests, either; he gave twelve. And rather than administer surveys and questionnaires, he put his subjects to work. In one task, for example, he instructed them to view fifteen cartoons and then come up with as many humorous captions as they could within thirty minutes. Knowing that the captions would vary in quality, he asked a team of independent observers to rate the wittiness and originality of each one.

Ruch's findings were clear—extraverted subjects produced the most humor. The more outgoing they were, the more captions they created. They were also the most cheerful, least serious, and most likely to self-report a sense of humor. In short, they were the most fun to be around.

The subjects who rated high on psychoticism proved quite different. They scored low on seriousness and produced fewer captions. However, they distinguished themselves by the quality of those captions. Specifically, their contributions were judged by the independent observers to be significantly funnier than everybody else's. So, being assertive, manipulative, and dogmatic might make you less prone to tell jokes, but at least the jokes will be funnier and more likely to make people laugh.

This finding helps explain why some studies find relationships between certain personality characteristics and humor, and others don't. It's not sufficient just to measure how many jokes a person tells, because quantity is very different from quality. We've all known people who love to tell jokes and amuse others around them. Sometimes they're funny, but other times they're simply annoying. I'm not saying that a person needs to be mentally unbalanced to crack a good joke, or that all good comedians are necessarily schizophrenic. Rather, my point is that when we "cultivate" conflict, both within our brains and with others around us, we're more likely to find our humorous side. By the same token, overactive brains aren't a bad thing, at least when it comes to humor. As we've seen, it's important to have a bit of edginess, just enough to make those jokes funny. Too little and we become boring. Too much and we become institutionalized.

One implication is that psychotic individuals (again, I'm not implying any pathology) are more likely than others to speak out in awkward or socially unacceptable ways to make a good joke. But we haven't yet considered the difference between being funny and having a good sense of humor. As we all know, there's a difference between telling a joke and being able to enjoy one when it's presented to us. It's a matter of production versus appreciation. Are certain people better at "getting jokes" than others?

The answer is "yes"—and here, too, the explanation has to do with overactive brains. There's another group of individuals who are highly tuned for appreciating humor, and it's because their minds are more active than those around them. We call these individuals sensation-seekers.

Sensation-seeking individuals can be thought of as combining all three of Eysenck's key personality traits. Like extraverts they're highly excitable, always seeking out new social situations. Yet they're not always sociable. True sensation-seekers don't care whether their actions are harmful to themselves or others—and if unchecked the result can be dangerous lifestyles and antisocial behavior. This makes them more than a little neurotic. And like psychotics, sensation-seekers frequently have high levels of testosterone, which itself is strongly correlated with drug use and sex. In a sense, then, sensation-seeking is like turning the dial of one's personality to ten and letting the chips fall as they may.

We know that sensation-seeking individuals are especially responsive to humor because we can look at their brains as they process jokes. Which brings us to the topic of absurd humor, the kind that doesn't lead to easily resolved punch lines (e.g., *What's yellow and can't swim? A bulldozer!*). Most of us respond to absurd humor with "quiet" brains, primarily because we're not sure what to make of it. But we see a different reaction among sensation-seekers. For them, it leads to *more* brain activation, because it's seen as a challenge. This form of humor allows them to work as hard as they want to get the joke, and since it's ridiculous, there's no boundary for how much exercise their brains can get. So if you want to know if any of your friends have a particularly keen sense of the absurd, show them the cartoon in Figure 4.1. An inactive, lazy brain will immediately give up trying to make sense of it. A sensation-seeker's brain will keep probing.

FIGURE 4.1. An example of absurd humor, as used in a study showing that sensation-seeking individuals experience high levels of brain activation when processing nonsense jokes. Reprinted from *Neuropsychologia*, Vol. 47, Andrea Sampson, Christian Hempelmann, Oswald Huber, and Stefan Zysset, "Neural Substrates of Incongruity-Resolution and Nonsense Humor," 1023–1033, 2009, with permission from Elsevier.

Other findings linking personality characteristics to humor appreciation are just strange or surprising. Herbert Lefcourt found that people with a strong sense of humor are also more environmentally conscious. Rod Martin and Nicholas Kuiper found that men with strong Type A characteristics, such as ambition and adherence to deadlines, enjoy jokes more than their laid-back colleagues do (but women showed no such difference). And then there are the truly strange studies, like the one titled, appropriately, "Humor and Anality." It tested Freud's theory that laughter is our way of dealing with sensitive topics, such as defecation. According to Freud, a lot in life is forbidden. Without getting into details, anality is one of those forbidden things, stemming from the need to discharge waste while also keeping ourselves clean and orderly. A person who feels the need to control everything is anal. And, according to the study, so is a person who particularly likes this joke:

> An agitated mother rushed into a drugstore, screaming that the infant in her arms had just swallowed a .22-caliber bullet. "What shall I do?" she cried. "Give him a bottle of castor oil," replied the druggist, "but don't point him at anybody."

The anality study looked for connections between preferences for jokes like this one and extreme organization, though I'm not going to discuss the results here, mostly because I find the topic silly. But as we can all agree, we humans are a complicated species, and we vary in many ways, including how much we enjoy jokes about poop. Ultimately, our sense of humor can be a great way of telling us apart. It can help us better understand ourselves and who we really are.

## THE FAIRER SEX

"It is axiomatic in middle-class American society that, first, women can't tell jokes—they are bound to ruin the punch line, they mix up the order of things, and so on. Moreover, they don't 'get' jokes. In short, women have no sense of humor."

In my job at the University of Maryland I'm surrounded by incredibly smart women. According to a 2007 survey by the National Institutes of Health, 43 percent of postdoctoral fellows in the biomedical sciences are women. In fields like mine, which includes psychology and sociology, that number is even higher. So it's almost wrong to call women a minority, at least in academia. Yet they're still often treated with less respect than men, paid less, and subjected to generalizations like the one above.

What's even more amazing is that the author of that quotation isn't a man but a woman, and a well-respected one at that—namely, Robin Lakoff, a prominent sociolinguist and feminist who frequently writes about language differences between the sexes. What Lakoff actually meant, though lost when taken out of the context, is that women communicate differently than men and, consequently, are often subjected to misunderstandings in male-dominated environments. Because their language tends to be powerless, they can't tell jokes, at least not effectively, and so are robbed of an important social function. The idea is controversial, though it does raise a good question: Are women less funny than men?

I have a hard time believing so, but this question does highlight several important differences between men and women, including how they communicate. Many such differences are subtle and difficult to recognize, but humor is not subtle. Humor is direct, and it can be useful for recognizing gender differences. If women really are less skilled at cracking jokes, what does that say about how they think?

One of the largest scientific studies on gender and humor was conducted by the psychologist and noted laughter researcher Robert Provine. Like Richard Wiseman (whom we met in Chapter 1), Provine wanted to examine humor in a natural setting. However, he had no interest in jokes. Rather, he wanted to see how men and women differ in terms of frequency of laughter. To do that, he sent assistants out to eavesdrop on people in public places. They listened to conversations at parties, took notes on subways, and monitored people ordering coffee in diners—all to collect what Provine calls "laugh episodes." Finally, after almost a year of collecting more than a thousand such events, Provine was finally able to say who laughed more in natural settings.

Women, he found, laughed more than men, up to 126 percent more. So, it certainly isn't true that women have no sense of humor. Women talking to other women generated the most laughter, accounting for 40 percent of the recorded episodes. Men talking to other men led to laughter only about half as often. In addition, women laughed more in mixed conversations (i.e., between men and women), and it didn't even matter who was speaking. Whether it was the man or the woman doing the talking, females were over twice as likely to laugh as their male counterparts.

These data reveal that women do indeed laugh and enjoy a good joke, though probably for different reasons than men. Laughter isn't offered easily among men. Perhaps it's a macho thing, or maybe they're by nature more reserved, but men are far likelier to elicit laughter from the person next to them than to laugh themselves. Put two women in a room and they'll soon share a laugh, but when genders are mixed, it's the men who are the clowns and the women who are the audience.

Perhaps this explains why women are less likely to go into professional comedy. In 1970, the percentage of professional female stand-up comics was approximately 2 percent. It rose to 20 percent in the 1990s and is now close to 35 percent, but this last figure may be deceptive. Shaun Briedbart, a comedian who has written for Jay Leno and appeared on the television show *The Last Comic Standing,* came up with that last number by counting the number of women performers at open mike nights in New York City, which is far from a professional setting. "The percentage of professional working comics is probably much lower . . . because it takes years to go from starting out to making money," noted Briedbart. "And maybe only one percent ever make it to the professional level."

Why do women struggle in the world of comedy? One way to find out is to look at the brains of comic artists and comedians. So far, we've seen that several areas of the brain are activated when we process humor, including those associated with conflict and reward. Yet, we haven't looked to see if that pattern is the same for everybody. Maybe men and women have different kinds of brains, and that's why they find different things funny.

Allan Reiss is a professor of psychiatry and behavioral sciences at Stanford University, and his interest in humor began with a simple question: What triggers cataplexy? A disease affecting about one in ten thousand Americans, cataplexy involves the occasional, sudden loss of voluntary muscle control. Although it's different from the epileptic fits described at the beginning of this chapter, its consequences can be just as troubling. Cataplectic incidents usually start with a slackening of the facial muscles, followed by weakness of the knees and legs. Muscles begin to tremble, speech starts to slur, and finally the entire body collapses. Then, the sufferer is left to wait, lying motionless yet completely alert, killing time until the spell ends. Reiss knew that many cataplectic incidents start as laughter, a fact that made him wonder why so little is known about the brain's emotional responses. To understand the disease, he would have to study what happens in our brains when we find something funny.

First, he had ten males and ten females view forty-two cartoons while being monitored using an MRI scanner, then he asked them to rate the funniness of each one on a scale from 1 to 10. Half the cartoons had previously been assessed as funny whereas the other half were not—a difference that Reiss hoped would allow him to compare brain responses based on joke quality. In addition, he made subtle changes to some of the cartoons, modifying them just enough to ruin the punch lines. "I was fascinated by what very small changes were necessary," he reported afterward. "Changing just one word in the caption could make the difference between a hilarious cartoon and a totally unfunny one."

As expected, Reiss found that both males and females showed strong brain activation in regions known to process visual images, as well as in frontal areas dealing with the logical mechanisms associated with humor. Men and women also scored similarly on the number of cartoons they found funny. In other ways, however, they differed substantially. For example, women showed significantly more activity in the left inferior frontal gyri, a region important for language. This region includes Broca's area, which is essential for producing words and speech.

Another subset of regions also showed more activation in women during humor processing—namely, the dopamine reward circuit. As discussed in Chapter 1, these are the regions that are responsible for giving us pleasure when we eat chocolate—or understand a joke. They were activated in both males and females during joke processing, but to a far greater degree in females. Such activation even increased for women the funnier they found the jokes. For men, activation remained moderate for all jokes, except the ones with the funny parts removed—which led to a *decrease* in activity.

"The results help explain previous findings suggesting women and men differ in how humor is used and appreciated," said Reiss in a press release distributed shortly after his paper's publication. The greater activation within language and reasoning centers of the frontal lobe suggests that the brain's analytical machinery becomes more intensively engaged in women than in men when reading jokes. This indicates either that women approach jokes with a more open mind, allowing their brains to ramp up once the joke begins, or that they dedicate more cognitive effort to coming up with a resolution when it's over. Reiss prefers the first interpretation: "This difference in brain activity seems to have more to do with [women's] expectations than their actual experiences. . . . Women appeared to have less expectation of a reward, which in this case was the punch line of the cartoon. So, when they got to the joke's punch line, they were more pleased about it."

This difference in expectations tells us a lot about how the two sexes look at life. Men expect a lot, and when they don't get it they become sour. Women expect little and are happy when they get anything at all. When they "get" the punch line, their reward centers light up because the pleasure is so surprising. Women don't laugh more than men because their brains are more active—they laugh more because their minds are more open.

Is it possible that women approach humor with a more open mind because men expect them to laugh at all their jokes? Or could it be that they laugh more because men give them so much reason to? Both explanations seem possible, but there's a third option, one that also helps clarify why women laugh more when men are present rather than absent—perhaps Lakoff was right when she claimed that women are more sensitive to humor because they're so commonly discriminated against. Laughter may be their only defense. Certainly no one can deny that humor often includes sexual biases.

Sexist jokes are an especially controversial issue, with so much already written about the topic it's difficult knowing where to start. For example, we know that women dislike jokes that make fun of female victims. We also know that they dislike sexual humor that objectifies their gender. My favorite finding, however, is that men like the cartoons from *Playboy* more than those from *The New Yorker*, whereas women express no such preference. Actually, that's an oversimplification, because the study looked at a lot more than just this, but it did find that men rate sexist cartoons from *Playboy* up to 25 percent funnier than those from more journalistic periodicals chosen for their "relative innocence."

By now, no one should be surprised that women aren't fans of sexist jokes. But this doesn't mean that they're the more sensitive gender. Consider, for instance, this joke from Wiseman's LaughLab experiment, a rare example of women laughing at men:

> *A husband stepped on one of those penny scales that tell you your fortune and weight and dropped in a coin. "Listen to this," he said to his wife, showing her a small, white card. "It says that I'm energetic, bright, resourceful, and a great person." "Yeah," his wife nodded, "and it has your weight wrong too."*

Only 10 percent of the men in Wiseman's experiment found that joke funny, about as low a rating as you can get. For women—well, it ranked much higher.

Nobody likes being laughed at—women and men alike. But there's a broader question regarding the impact of sexist jokes on our behavior: Do sexist jokes reflect gender biases, or do they create them?

Psychology has well established that stereotypes have strong, negative impacts on our beliefs. Studies have shown, for example, that people who see African Americans portrayed in stereotypically negative roles in comedy skits are quick to adopt negative attitudes toward that group in real life. Exposure to such stereotypes can even increase the likelihood of falsely accusing African Americans of committing a fictional crime.

Sexist humor has similar impacts on perceptions of women, according to a study of sexist attitudes conducted by Thomas Ford at Western Carolina University. Ford first gave groups of adult males assessments of existing sexist beliefs, asking them to agree or disagree with statements like "Women seek to gain power by getting control over men." From those assessments, each subject was classified as possessing either low or high hostile sexism. Next, some of the subjects read a series of sexist jokes targeting women (e.g., *How can you tell that a blonde has been using the computer? There's Wite-Out on the screen!*) along with equally aggressive jokes not targeting women (e.g., *What's the difference between a golfer and a skydiver? A golfer goes whack . . . damn, a skydiver goes damn . . . whack*). As a comparison, other subjects read a series of nonsexist and sexist stories not involving humor.

To see what impact the sexist jokes and stories had on subject attitudes, Ford described the National Council of Women, an organization committed to the political and social advancement of women and women's issues, and asked all of the men to imagine making a donation to this organization, up to $20. They didn't have to commit any actual money, only to imagine themselves doing so. The final amount they chose to give was what Ford regarded as his dependent measure.

When he analyzed his data without taking into consideration the subjects' existing sexist beliefs, the jokes appeared to have no impact on how much money they committed to the organization. However, when he differentiated the responses of those scoring low and high on the sexist scale, a very different picture emerged.

Ford found that, compared to low-sexist subjects, high-sexist subjects were willing to commit much less money to the National Council of Women—but only after reading the sexist jokes. The nonsexist jokes, as well as the nonhumorous sexist stories, had no impact on their donations. To confirm his findings, Ford varied his experimental design by asking the subjects how much money a fictional university should cut from student organizations with similar woman-related causes. The results were the same. High-sexist subjects advocated the most drastic cuts, but only after reading the sexist jokes.

If you're like me, you find these results surprising and even a little frightening. Sexist humor does indeed appear to be more insidious than misogynist propaganda. It could even be that humor elicits opinions and emotions more effectively than direct prejudice because it works at a level below conscious awareness. In other words, by "flying below the radar," humor amplifies existing prejudicial beliefs, giving them a voice without allowing them to be openly questioned.

Because it reveals how influential humor can be, Ford's research is a good argument against stereotype-driven humor—even lawyer jokes. Granted, it only matters if we already have prejudicial attitudes toward these groups (in fact, low-sexist subjects pledged *more* money to the National Council of Women after the sexist jokes). But as we've seen, humor always contains two messages: what the humorist is saying, and all the other stuff left unspoken. When that unspoken stuff is hurtful or prejudicial, the easiest way to slip it in is to use a joke. Again, it's a matter of intent.

## SPECIALIZATION IS FOR INSECTS

Near the beginning of this chapter I talked about specialization and how we humans evolved to our current successful position by being extreme generalists. It's time to return to that topic, which brings me to one of my favorite quotations of all time:

A human being should be able to change a diaper, plan an invasion, butcher a hog, conn a ship, design a building, write a sonnet,
balance accounts, build a wall, set a bone, comfort the dying, take
orders, give orders, cooperate, act alone, solve equations, analyze a
new problem, pitch manure, program a computer, cook a tasty meal,
fight efficiently, die gallantly. Specialization is for insects.

This quote appears in Robert Heinlein's *Time Enough for Love,* a
science-fiction novel about a two-thousand-year-old man who lives so
long he loses the will to go on. So far in my life I've accomplished
thirteen of the items on Heinlein's list, and unless I spend some time
on a farm or contract a deadly disease, I'm unlikely to complete many
more. I like the quote because it shows just how varied life can be, and
how wonderfully flexible our brains have become in preparing us for
life's different challenges.

Here's another, perhaps more familiar quote by Heinlein: "When
apes learn to laugh, they'll be people." I like this one too, because it
implies that laughter is part of what makes us human. In the pages that
follow, we'll put this theory to the test, not by determining whether
apes have the capacity to laugh—they do, as we've already seen—but
by looking at the ways that our complex human brains develop, cul-
minating in an ability to laugh. Few people other than Aristotle would
argue that an infant who hasn't yet laughed doesn't have a soul, but
I think we'd all agree that people of different ages laugh at different
things. This variability says a lot about our cognitive development, as
well as about how complex and "human" our brains have become.

Consider, for example, one of our first developmental hurdles: "object permanence." This is the ability to recognize that the world exists
separate from our perceptions, and that when we close our eyes, the
world doesn't disappear. It takes infants up to two years to fully appreciate this fact, which is why toddlers love to play peekaboo. There's a
time in our development when seeing something disappear means that
it's gone forever. Sometime later we recognize that objects and people
continue to exist even when they can't be seen. Between these periods

there's a transitional phase when the brain experiences conflict—a moment of confusion or indecision. A child who no longer enjoys peekaboo probably has mastered the concept of object permanence. One who is terrified by this game probably hasn't figured out the trick.

As it turns out, apes not only laugh, they have a pretty firm grasp on object permanence too. As do dogs, cats, and a few species of birds, including the raven. For example, if you hide food behind a barrier, then move the barrier around, each of these animals will recognize that the food is still there, even after an extended period of time. Have you ever read about scientists claiming that dogs are smarter than cats? Tests involving object permanence are how they make such claims, because dogs perform slightly better than cats on such tests. So do ravens, so score one for the birds.

Examining humor in children allows us to see what cognitive stage of development they're in. After object permanence, a big challenge for children is achieving "theory of mind"—the ability to attribute mental states to others and to understand that others have beliefs and intentions that are different from our own. In short, it's the ability to overcome egocentrism.

Children below the age of about six can't tell the difference between a lie and a joke because they lack the theory of mind to recognize that these are different things. For similar reasons, they also don't understand irony and sarcasm. In each of these cases the literal message is different from the intended one, and the listener must recognize this by considering the motivations and intentions of the speaker. Because children younger than six typically don't get that someone can have different intentions than their own, the humor in sarcastic statements is lost. One study found that many children as old as thirteen fail to recognize sarcasm in spoken remarks, even when they realize that the remarks themselves are incorrect.

I have never changed a diaper, number one on Heinlein's list, and I've also never raised a child. But I have many friends who have, and they consistently claim that it's a cruel fate seeing them master the art of sarcasm just in time to become a hormone-ridden teenager.

One of the last major challenges for children is "operational thinking," the ability to reason abstractly. At early ages we learn to manipulate objects in our environment, and even organize and classify them. Eventually we learn to use symbols for these objects, and when we get really advanced we do the same for things we can't see or touch, like numbers. Children who call a pet dog a "cat" or say "hi" to a person in a photograph—both are attempts at humor—are essentially breaking newly learned rules about abstract names referring to concrete things. They're playing with the fact that the representations are different than the objects themselves.

Development doesn't end in childhood, of course. It extends throughout the life span, meaning that humor preferences change later in life too. As most of us have learned through personal experience, one key aspect of getting older is that we lose cognitive flexibility. It becomes harder to learn new things and to approach new situations with open, flexible minds. Another consequence is that we stop caring what other people think, which can have a sizable impact on our sense of humor.

To explore why this is the case, let's consider another study conducted by the German psychologist Willibald Ruch. This one was his biggest yet, examining more than four thousand subjects ranging from age fourteen to sixty-six. Ruch started by giving his subjects a sense-of-humor questionnaire, which divided humor into two types: "incongruity humor," which involves the traditional surprise and resolution stages described earlier, and "nonsense humor," which also involves incongruity but leaves the resolution stage unresolved for the sake of the ridiculous. We've seen this kind of joke already: *Why did the elephant sit on the marshmallow? Because she didn't want to fall in the hot chocolate.*

After measuring the subjects' preferences for these two humor types and administering additional personality assessments, Ruch analyzed the data to determine whether these preferences changed with age. They did. Not surprisingly, he found that as people get older, they like nonsense humor less and incongruity humor more—probably because, by a certain age, we all come to expect things to make sense.

But the most interesting finding emerged when Ruch compared these results with conservatism, which he'd also assessed. Conservatism is a difficult thing to measure, as you might expect, so Ruch was forced to create his own test. Composed of several questions from other personality assessments asking questions about traditional family ideology, liberal upbringing of children, and orientation toward work, his test measured how averse the subjects were to change and how traditional they were in their social outlook. Ruch found that age differences in humor are strongly correlated with conservatism. The more people disliked nonsense humor, the more conservative were their beliefs.

This effect was quite strong, accounting for 90 percent of the variance in incongruity-humor liking and 75 percent for nonsense-humor disliking. In fact, it was strong enough to suggest that taste in humor is largely driven by conservatism alone.

Before writing this book I would never have guessed that our brains have an optimal age for humor. It has been said that children are fools if they are not liberals, just as adults are fools if they are not conservatives—and this may very well be true, at least in terms of brain plasticity. Young brains are flexible and open, leading to an affinity for liberalism and elephant jokes. Conflict is less of a problem for children than for adults because it helps them grow and learn. But as we get older, our perspectives alter. Change becomes less welcome, as does absurdity, and learning becomes less important than making things fit. It's not a happy thought, at least for those of us in that second group, but it's an important one to recognize.

Indeed, by revealing so much about ourselves, humor may be the best way of learning who we really are. It's an intriguing idea, one that will get further attention in the next chapter. Except, next we won't be singling out women, children, or conservative adults. Instead, we'll be looking at individuals who don't have any brains at all.

# 5

## » OUR COMPUTER OVERLORDS

*The question of whether computers can think is
like the question of whether submarines can swim.*
—EDSGER W. DIJKSTRA

"THIS WAS TO BE AN AWAY GAME FOR HUMANITY." SO SPOKE
Ken Jennings, author, software engineer, and holder of the longest
winning streak on the television show *Jeopardy!* He had been invited
by producers of the show to compete against, of all things, a computer,
which IBM had developed as part of its artificial intelligence research
program. It seemed like an intriguing idea, at least until he entered the
auditorium where he would compete and saw that the entire crowd
was against him. Rather than filming in its usual Los Angeles location,
the show had been transported to Westchester County, New York, the
site of IBM's research labs. As soon as the lights went on, the audience
cheered. But they weren't cheering for their own species. They were
rooting for the competition.

"It was an all-IBM crowd: programmers, executives. Stockholders all!" said Jennings. "They wanted human blood. It was gladiatorial out there."

The challenge was daunting; Watson was a marvel of engineering, and everyone knew it. Built from ninety clustered IBM Power 750 servers and running thirty-two massively parallel Power7 processors, Watson was capable of holding more than 16 terabytes of memory. That's a 16 with twelve zeros after it. And it operated at more than 80 teraflops, which meant that it could perform 80 trillion operations—per second. In short, it was built to hold its own against whatever combination of water, salt, and proteins its competitors threw at it.

Despite Watson's power, the historical advantage still belonged to humans. IBM had developed Watson to compete on *Jeopardy!* because this is exactly the arena where computers typically fail. Watson may have had incredible computing power, but the game of *Jeopardy!*—like life—is messy. Winning takes not just real-world knowledge but the ability to recognize irony, slang, puns, pop-culture references, and all sorts of other complexities. But it also requires knowing what you don't know. In other words, you can't just guess at every opportunity, because penalties for errors add up.

Consider, for instance, the sentence "I never said she stole my money," which the IBM engineers offered as an example of the kind of ambiguity for which humans are specialized. There are literally seven different meanings those words can convey, an impressive number given that the sentence contains only seven words. If you don't believe me, read it out loud yourself, each time emphasizing a different word. All it takes is an inflection here or a change of stress there, and the entire intention is changed. Recognizing this kind of ambiguity is something humans do with ease, but computers—well, let's just say that computers don't like to be confused.

After the first day of competition Jennings performed relatively well against both Watson and Brad Rutter, the other human contestant. At one point Rutter, who owned the distinction of having won the most money in the show's history, was tied with Watson at $5,000. Jennings had $2,000. Then things got out of hand.

Rather than being stumped by vague or confusing clues, Watson thrived on them. It knew that "The ancient Lion of Nimrud went missing from this city's national museum in 2003" meant "Baghdad" and that "An etude is a composition that explores a technical musical problem; the name is French for this" referred to "study." Granted, it also made mistakes—for example, when it gave "Toronto" as an answer for the category of "US cities." But the gaffes were minimal and Watson won handily with $35,734, compared to Rutter's $10,400 and Jennings's $4,800.

The second match, which was to be aired on the final night of the competition, February 16, 2011, removed any question as to who the new *Jeopardy!* champion would be. By the time the contestants reached the last round, which always ends with a wager on a single, final clue, Watson had a significant lead. The final question, "What novel was inspired by William Wilkinson's *An Account of the Principalities of Wallachia and Moldavia*," was answered correctly by all three contestants (Bram Stoker's *Dracula*), but it didn't matter. Watson had already won the match, though Jennings had one last surprise inside him.

Below his final answer he wrote: "I, for one, welcome our new computer overlords."

This was a play on a classic line from an episode of *The Simpsons* in which a clueless news anchor, believing that the earth has been taken over by a master race of giant space ants, decides to suck up to his new bosses. "I, for one, welcome our new insect overlords," he says. "And I'd like to remind them that as a trusted TV personality, I can be helpful in rounding up others to toil in their underground sugar caves."

Jennings may have lost the match, but he won several hearts—especially when he took an extra jab at the computer that had just defeated him: "Watson has lots in common with a top-ranked *Jeopardy!* player. It's very smart, very fast, speaks in an even monotone, and has never known the touch of a woman."

In fact, Jennings did something that Watson could never do—he cracked a joke. Using its massive computing power Watson was able to overcome the problem of ambiguity, but it couldn't tell a joke because jokes require not just recognizing ambiguity but exploiting it too. That's a lot to ask, even from such a powerful machine as Watson.

In today's world, there's almost nothing computers can't do. They help to fly our planes, drive our cars, and even give medical diagnoses. One of the last things we thought computers could do is deal with ambiguity like humans—which is why Watson's accomplishment was so impressive. Contrast this with Deep Blue's defeat of chess grandmaster Garry Kasparov in May 1997. Deep Blue was capable of examining 200 million chess moves in the three minutes allocated for each move, but it didn't have to deal with messy things like language. Chess, though complex, is still a well-defined problem: there's never any doubt regarding the purpose of the game or what the next potential moves are.

Yet, both Watson and Deep Blue highlight the importance of flexible thinking. That flexibility was seen in Watson's ability to interpret subtle meanings and make reasonable guesses as to possible linguistic interpretations, and also in Deep Blue's surprising chess moves. Flexibility is key, especially for activities that computers typically struggle with, such as being creative. Writing sonnets, composing symphonies, telling jokes—these are things computers will never be able to do. Or will they?

Consider Game Two of the 1997 match between Kasparov and Deep Blue, which ended with a win for the computer. About thirty moves into the match, Kasparov realized he was in trouble and decided to sacrifice a pawn. Taking that pawn would have given Deep Blue a distinct advantage. Every chess-playing program ever created would have taken it, and so would most chess masters. There were no obvious drawbacks to the move. Yet Deep Blue rejected the bait. Instead, it moved its queen to "b6," a position with less immediate benefit. But it also disrupted Kasparov's attempt for a comeback—a ploy that shocked Kasparov so deeply that he claimed humans must have intervened.

There was no way a computer, or anyone less than a grand master, could have seen what he was planning and countered so effectively. Computers simply aren't that creative.

While writing this book I found it fascinating to learn that computer chess programs hold an advantage over humans only when contestants are given *less* time to ponder moves, not more. This seems counterintuitive—if given unlimited time to search possible moves, computers should be stronger players than humans, not worse. With so much computing power, the extra time should be a benefit. But it isn't. Why? For the same reason that computers can't tell good jokes. They're not messy thinkers. They seek solutions linearly, rather than letting their minds argue and drift until some solution pops out of nowhere. All the time in the world doesn't help if you don't know how to look.

Over the previous four chapters, we've seen that our messy thinking has some benefits. One is humor. Messy thinking also helps with chess and *Jeopardy!* because it allows us to search vast arrays of possible moves holistically, using intuition rather than algorithms. In each of these cases, the goal isn't to derive some simple solution. It's to make unexpected associations, and even connect ideas that have never been connected before.

All this is a roundabout way of saying that despite Watson's victory, humans are still the only ones who are truly creative. Yet, scientists are making great strides in the field of computer intelligence, and in this chapter we'll see how. We'll explore the complex and mysterious nature of creativity, discovering how humor provides unique insights into what that really is. And we'll see what all this has to do with telling jokes—and why perhaps computers aren't as far away from being funny as we might think.

## PATTERN DETECTION AND HYPOTHESIS GENERATION

*What kind of murderer has moral fiber? A cereal killer.*

I know this joke isn't particularly funny, but what if I told you it wasn't written by a person? What if I told you it was written by a computer?

The cereal killer joke was just one of many jokes created by a program that you can operate yourself online. Just visit the University of Aberdeen's website and look for a project called The Joking Computer. The program will ask you to choose a word to start with—this is the nucleus around which your new joke will be formed. Then it will ask a few more questions, such as what words rhyme with the one you chose. And, finally, it will show the completed joke. When I logged on and tried it for myself, I came up with the following zinger: *What do you call a witty rabbit? A funny bunny.*

Again, not terribly funny, but when the cereal killer joke was submitted to Richard Wiseman's LaughLab competition, it actually outperformed many of the human-created ones. It didn't win—it didn't even come close, ranking just below the middle of the pack—but it also didn't stand out as odd or incomprehensible. That in itself is quite an accomplishment.

These two jokes show how simple joke construction can be. But they're only mildly funny because they rely on a simple word trick without much surprise. One could even argue that they aren't creative, because they're so simple. The computer just picks a word, then looks for synonyms and rhymes until finally it comes up with a solution. There's not much thinking involved, so to what extent can such programs reveal how humans really think?

Quite a bit, as it turns out. Creative behavior can be as simple as combining old ideas in new ways. And, as we learned in previous chapters, jokes are funny because they force us to confront mistakes of thinking—for example, errors in scripts. When we create a joke we're not inventing new thoughts or scripts, we're connecting ideas in new ways.

"Humor is essentially a matter of combinatorial creativity," says Margaret Boden, cognitive scientist, professor of informatics at the University of Sussex, and author of *The Creative Mind: Myths and Mechanisms*. "Elephant jokes, changing light bulb jokes: those are two styles which are easy to recognize. All you need is the ability to connect ideas in novel ways and you have yourself a joke. Deciding why one joke is funnier than another one—well, that's another matter."

This is the classic issue in computer science: computers find it easy to create new things but nearly impossible to assess their usefulness or novelty. This failing is most obvious in the realm of humor, because knowing how funny a joke is takes world knowledge—something that most computers lack, even Watson. Consider, for example, a joke made by The Joking Computer's successor, the Joke Analysis Production Machine (JAPE): *What kind of device has wings? An airplane hangar.* The reason JAPE thought this joke was funny was that it classified *hangars* both as places for storing aircraft and as devices for hanging clothes. That's accurate (to the extent that we accept the misspelling of *hangers*), but most humans know that a long piece of wire holding a shirt isn't much of a "device."

Even though it followed its formula correctly, JAPE was unsuccessful specifically because it failed to recognize the lack of humor in the final product. This challenge might also explain why there are so many joke production programs but so few specialized for joke recognition. To write a joke, all you need is a strategy, such as manipulation of rhymes or replacement of words with synonyms. That's the tool used by the online program Hahacronym, which uses a stored database of potential replacements to identify funny alterations of existing acronyms. *What does FBI stand for? Fantastic Bureau of Intimidation. MIT? Mythical Institute of Theology.*

Of course, identifying good humor requires more than simple tricks, since there are no shortcuts for classifying the myriad ways to make a joke. Typically, humor recognition programs meet this challenge through massive computing power, like Watson did when answering *Jeopardy!* questions. Such programs look for language patterns, especially contradictions and incongruities. In this sense they're pattern detectors. But to be effective, they must access vast amounts of material—as in, millions of pieces of text. (As a comparison, since starting this book you've read about forty thousand words yourself.)

One example of a pattern detection program is Double Entendre via Noun Transfer, also known as DEviaNT. Developed by Chloé Kiddon and Yuriy Brun at the University of Washington in Seattle, it identifies words in natural speech that have the potential for both sexual and nonsexual meanings. Specifically, it searches text and inserts the phrase "That's What She Said" during instances of double entendres (a task of great practical importance to frat houses and fans of *The Office*). DEviaNT is distinctive in that it's not just a joke creator but a humor recognition program too, because it takes a sense of humor to know when to "interrupt."

DEviaNT was first taught to recognize the seventy-six nouns most commonly used in sexual contexts, with special attention to the sixty-one best candidates for euphemisms. Then it read more than a million sentences from an erotica database, as well as tens of thousands of nonerotic sentences. Each word in these sentences was assigned a "sexiness" value, which, in turn, was entered into an algorithm that differentiated the erotic versus nonerotic sentences. As a test, the model was later exposed to a huge library of quotes, racy stories, and text messages as well as user-submitted "That's What She Said" jokes. The goal was to identify instances of potential double entendre—a particularly interesting challenge, noted the authors, because DEviaNT hadn't actually been taught what a double entendre was. It had been given only lots of single entendres, and then was trained to have a dirty mind.

The researchers were quite pleased when DEviaNT recognized most of the double entendres it was presented, plus two phrases from the nonerotic sentences that had acquired sexual innuendo completely by accident ("Yes, give me all the cream and he's gone" and "Yeah, but his hole really smells sometimes"). DEviaNT's high degree of accuracy was especially impressive given that most of the language it was tested on wasn't sexual. In effect, it was trying to spot needles in haystacks.

But that's cheating, you might claim. DEviaNT didn't actually understand the sexual nature of the jokes. It didn't even know what it was reading. All it did was look for language patterns, and a very specific type at that. True, but these arguments also assume that "understanding" involves some special mental state, in addition to coming up with the right answer. (Or, when recognizing bawdy jokes, knowing when to exclaim "That's what she said!") As we'll soon see, that's a human-centric perspective. Maybe we underestimate computers because we assume too much about how they should think. To explore that possibility, let's turn to one last computer humor program—the University of North Texas's one-liner program, developed by the computer scientist Rada Mihalcea.

Like DEviaNT, this program was trained to recognize humor by reading vast amounts of humorous and nonhumorous material. Specifically, it was shown sixteen thousand humorous "one-liners" that had been culled from a variety of websites, along with an equal number of nonhumorous sentences taken from other public databases. Mihalcea's goal was to teach the program to distinguish between the humorous sentences and the nonhumorous ones. But the program had two versions. One version looked for certain features previously established as common in jokes, such as alliteration, slang, and the proximity of antonyms. The second version was given no such hints at all and simply allowed the program to learn on its own from thousands of labeled examples. After training, both versions were shown new sentences and asked to identify which were jokes and which weren't.

Mihalcea was surprised to see that the trained version of the program, the one told which features are most common in jokes, did relatively poorly. Its accuracy hovered only slightly above chance at recognizing humor, meaning that the hints weren't very helpful. By contrast, the version that learned on its own—using algorithms such as Naive Bayes and Single Vector Classifier, which start with no previous knowledge at all—reached accuracy levels averaging 85 percent. This is a fairly impressive outcome, especially considering that many humans also have difficulty recognizing jokes, especially one-liners.

Mihalcea's finding is important because it shows that imposing our own rules on computers' thinking seldom works. Computers must be allowed to "think messy," just like people, by wandering into new thoughts or discoveries. For humans this requires a brain, but for computers it requires an algorithm capable of identifying broad patterns. This is essential not just for creating and recognizing jokes but for all artistic endeavors. Watson needed to be creative, too. The programmers at IBM didn't try to define what problem-solving strategies Watson used to win at *Jeopardy!* Rather, they allowed it to learn and to look for patterns on its own, so that it could be a flexible learner just like the human brain.

Some people may argue that people aren't pattern detectors, at least not like computers. If you believe this, you're not alone. You're also wrong. Recognizing patterns is exactly how the human brain operates. Consider the following example: "He's so modest he pulls down the shade to change his ___." What's the first word that comes to mind when you read this sentence? If you're in a humorous mood, you might think of *mind,* which is the traditional punch line to the joke. If you're not, you might say *clothes.* Or maybe *pants.*

I share this joke because it illustrates how the human brain, like a computer, is a pattern detector. *Cloze probability* is the term that linguists use to describe how well a word "fills in the blank," based on common language use. To measure cloze probability, linguists study huge databases of text, determining the frequency at which specific words appear within certain contexts. For example, linguists know that the word *change* most often refers to replacement of a material object, such as clothes. In fact, there's a cloze probability of 42 percent that the word *clothes* would appear in the context set up by our example—which is why it was probably the first word you thought of. *Change* referring to an immaterial object, such as a mind, is much less likely—closer to 6 percent.

These probabilities have a lot to do with humor because, as already discussed, humor requires surprise, which in this case is the difference between 42 percent and 6 percent. Our brains, much like computers, do rapid calculations every time we read a sentence, often jumping ahead and making inferences based on cloze probability. Thus, when we arrive at a punch line like *mind,* a sudden change in scripts is required. The new script is much less expected than *clothes*, and so the resolution makes us laugh. Computer humor recognition works the same way, looking for patterns while also identifying the potential for those patterns to be violated.

Why, then, aren't computers better at cracking jokes than humans? Because they don't have the world knowledge to know which low-probability answer is funniest. In our current example, *mind* is clearly the funniest possible ending. But *jacket* has a low cloze probability too. In fact, the probability that people will refer to *changing their jacket* is about 3 percent—half the probability they'll talk about *changing their mind.* Why is the second phrase funny whereas the first one isn't? Because, with our vast world knowledge, people understand that changing our mind isn't something that can be seen through a window.

We know this because we've all stood in front of windows. No computer has ever stood in front of a window.

To understand why computers struggle when recognizing good jokes, think back to the EEG findings from Chapter 2. As we learned, our brains elicit two kinds of reactions to jokes—the P300 and the N400. The P300 reflects an orienting reflex, a shift in attention telling us that we've just seen something new or unexpected. The N400 is more semantic in nature. It measures how satisfying the new punch line is, and how well it activates a new perspective or script.

In that earlier chapter we also discovered that whereas all jokes elicit a P300, only funny ones elicit an N400, because these bring about a satisfying resolution. A related finding is that a word's cloze probability is inversely proportional to the size of the N400 it produces—the higher the cloze probability (i.e., the more we expect to see that word), the smaller the N400. This size difference reflects how easily new words are integrated into already constructed meanings, with easier integration meaning smaller N400s. At first you might think that cloze probability should influence the "surprise" response of the P300, but this isn't the case. Low-probability words aren't shocking, only incongruent. It's a matter of context—larger N400 responses mean that contexts are being shifted, while P300 responses mean that we're simply shocked, context having nothing to do with it.

It's a subtle difference, one that computers struggle with. To computers, there's no such thing as context, only a constant stream of probabilities. That's where we humans distinguish ourselves, bringing us back to the *constructing, reckoning,* and *resolving* stages from Chapter 2. The human brain doesn't just recognize cloze probabilities, it builds hypotheses and revises those hypotheses based on new evidence. It's always looking for patterns and constructing contexts, and by relying on both probabilities and expectations, it becomes an active manipulator of its environment rather than a passive receiver.

To see how this relates to humor, let's review a study conducted by the cognitive scientist Seana Coulson of the University of California at San Diego. Coulson's aim was to understand the human brain's sensitivity to both context and cloze probability. First, she showed subjects sixty sentences, some of which ended in a funny punch line and some of which didn't (e.g., "She read so much about the bad effects of smoking she decided she'd have to give up *the habit/reading*"). Only the joke endings were expected to bring about shifts in perspective. Next, she varied the cloze probability of the sentence endings, dividing them into two categories. Sentences for which the joke setup activated a salient, high cloze-probability ending—as in the above example—were labeled "high constraint." Those with a lower cloze-probability ending were called "low constraint." For example, "Statistics indicate that Americans spend eighty million a year on games of chance, mostly *dice/weddings*" is a low-constraint sentence because there are many possible endings— *dice* being only one of several low cloze-probability alternatives.

Not surprisingly, the N400s were bigger for sentences with funny punch lines than for those with unfunny ones. But this difference appeared only among the high-constraint sentences. That's because these were instances in which the subjects' world knowledge had set up some expectation and context, and the punch line brought a new way of thinking. Cloze probability is important to humor, but so is violation of our expectations. We're pattern detectors, but we're *constructors, reckoners,* and *resolvers* too. Computers' inability to incorporate all three processes is what causes them to struggle.

Before moving on to the next section, let's take one more look at how our thinking differs from a computer's. A little later we'll be addressing creativity, and how humor is just one example of this unique skill, a skill we still hold over our computer overlords. But for now, I want to drive home the point that the human brain is much more than just a parallel processor, or dozens of parallel processors linked together, as with IBM's Deep Blue or Watson. Indeed, it's like a child who can't sit still, always looking around the corner for what's coming next.

One benefit of computers is that they always follow directions: at any given time, we can tell a computer to stop working and tell us what it knows. It won't ignore our command, and it won't keep working and hope we don't notice. Humans are a different story. Our brains work so fast, and in such hidden ways, that it's nearly impossible to see what calculations they're really making. Analyzing jokes is especially difficult, because comprehension occurs in seconds. There's no way to stop people halfway through a joke and identify what they're thinking. Or is there?

"Semantic priming" studies are among the oldest in the field of psychology. The process is relatively simple: subjects are given a task— say, reading a joke—and then interrupted with an entirely different task that indirectly measures their hidden thoughts. For example, after reading the setup to a joke, they may be shown a string of letters and asked if those letters constitute a real word or not (called a "lexical decision" task). Imagine that you're a voluntary participant in a study and are instructed to read the following: "A woman walks into a bar with a duck on a leash. . . . " Then, the letters *S-O-W* appear on the screen and you're asked whether they form a real word or not. How long would it take you to recognize that *S-O-W* refers to a female pig?

Now, imagine that you're given the same task after reading the full joke: *A woman walks into a bar with a duck on a leash. The bartender says, "Where did you get the pig?" The woman says, "That's not a pig. It's a duck!" The bartender replies, "I was talking to the duck."*

Would you immediately recognize the meaning of *S-O-W* this time? Of course you would, because the word *pig* would have been activated in your mind. Without priming, it usually takes subjects between a third of a second and three times that long to recognize a given word. With priming (e.g., reading the above joke), that reaction time is decreased by a quarter of a second. This may not seem like much, but in the world of psychology it's a huge effect.

I mention semantic priming because Jyotsna Vaid, a psychologist at Texas A&M University, used this very task to find out the precise point at which subjects revised their interpretations and "got" a joke. For our example joke there are at least two possible interpretations. One of these is that the woman owns a pet duck and that the bartender doesn't know his birds from his boars. A good way to check for this interpretation is to use *P-E-T* in the lexical decision task, because if it's what subjects are thinking, then the word *pet* should be at the top of their minds. The second possible interpretation is that ducks can understand questions from surly bartenders, and that the woman is as ugly as a pig. For that one, *S-O-W* should be highly activated.

Earlier I noted that jokes become funny when scripts suddenly change due to an incongruous punch line—for example, a doctor's wife inviting a raspy-voiced man inside for an afternoon tryst rather than a chest exam. Now we're seeing the exact point at which these shifts occur. Not surprisingly, Vaid saw that the initial, literal interpretations of the jokes were dominant when subjects started reading. In other words, they had no choice but to assume the woman owned a pet duck. However, as soon as the punch line came and an incongruity was detected, the second interpretation became active too. The first one didn't disappear, though. Instead, it stayed active until the end of the joke, after the subjects had been given a chance to laugh. Only then did they make up their minds and move on—and the word *pet* stopped receiving facilitation in the lexical decision task. From these results we see that our brains build hypotheses, sometimes more than one at a time, and only as more evidence becomes available are old ones jettisoned like rotten fruit.

In a sense, then, we're built to be pattern detectors, always taking in new information and building stories. Much of the time those interpretations are correct. Sometimes they aren't.

And when they aren't, occasionally we laugh.

## TRANSFORMATIONAL CREATIVITY

"Computers are creative all the time," says Margaret Boden. But will they ever generate ideas—or jokes—that convince us they're truly creative without seeming artificial or mechanical? "Many respectable ideas have been generated by computers which have amazed us and that we value. But what we haven't seen is a computer that creates something amazing and then says, 'Don't you think this is interesting? This is valuable.' There are many systems which come up with amazingly novel ideas, but if there's any value in it, humans still need to persuade us why."

Boden is referring to a major problem with creativity—and a big challenge for humor researchers too. Creativity is subjective. Knowing when a punch line works or not, as with a painting or a sonata, requires being able to assess its value and novelty. But this capability is something many people lack, so imagine how difficult it must be for computers. How do we justify any work of art? How do we know that the punch line *An airplane hangar* isn't funny but a telegram-sending dog proclaiming *"But that would make no sense at all"* is?

According to Boden, there exists more than one type of creativity. In fact, there are several. The first and simplest form is "combinatorial creativity," which is the type displayed by simple programs like The Joking Computer. Combinatorial creativity involves combining familiar ideas in an unfamiliar way, as when words are put together to form a pun or rhyme. A good example, though it's not particularly funny, is the earlier punch line *A funny bunny*. Odds are that you never heard that joke before. It's possible that nobody has. But it didn't change the way you looked at jokes because it only manipulated a simple rhyme.

A second type is "exploratory creativity," which involves making new connections within existing knowledge. It's similar to combinatorial creativity, except that now we're dealing with a greater degree of novelty. Although outside the realm of humor, consider Paul McCartney's song "Yesterday." It wasn't the first Beatles ballad. It also wasn't the first recorded use of a cello, as classical musicians had been using the instrument for centuries. It was, however, the first modern rock song to give the cello such a prominent role. Now hip-hop artists such as Rihanna and Ne-Yo use it all the time.

Exploratory creativity allows us to make connections we've not seen before. Consider, for example, the Steven Wright joke *There was a power outage at a department store yesterday and twenty people were trapped on the escalator.* It's essentially an analogy, since elevators are different from escalators in their ability to trap people, thus triggering the script that Americans are overweight, lazy mall-dwellers. Probably no other comic has made the connection between escalator failures and sedentary shoppers, but Wright did and he got a pretty good joke out of it.

The third type of creativity, "transformational creativity," is something entirely different. It occurs when we're forced to restructure our thinking, and Boden cites post-Renaissance Western music as a salient example. Prior to the work of Austrian composer Arnold Schoenberg, orchestral music always had a tonal key. Composers sometimes introduced modulations in the middle of a piece, but they always returned to the original key by the end, signaling the work's theme. These modulations were often surprising, but they weren't transformational in the sense I mean here. A transformational change came only when Schoenberg created a new kind of music never heard before—"atonality." Though disturbing to many at first, Shoenberg's dropping of the tonal key was quickly adopted by others and then subjected to several exploratory alterations itself.

We see such variation in humor, too. Stand-up comedians approach their art in different ways, and this variety is what makes comedy clubs so fun. But not all comedians rewrite their genre. Jerry Seinfeld, although funny and fantastically successful at pointing out the obvious, didn't force us to look at comedy differently. Neither did Steve Martin, even though he's one of the smartest comedians ever to have graced the stage. Andy Kaufman, on the other hand, was a transformational creative genius. He created alter egos so believable that his audience didn't know if they were a joke or real. He pretended to get into fights with fellow actors and comedians during live performances—sometimes even storming off the stage. Once, he ended a performance by taking the entire audience out for milk and cookies.

Nobody had ever created comedy like Kaufman, just as nobody had told dirty jokes and offended audiences like Lenny Bruce. For every hundred Seinfelds or Martins, there's only a handful of Kaufmans or Bruces.

Returning to the brain for a moment, it's worth noting that no single brain region is responsible for this type of creativity. One scientific review of seventy-two recent experiments revealed that no single brain region is consistently active during creative behavior. There is, however, something special about people who make novel connections or imagine the unimaginable. What sets them apart is the connectivity within their resting brains. This finding was discovered by a team of researchers from Tohoku in Japan, who observed that people with highly connected brains—as measured by shared brain activity over multiple regions—are more flexible and adaptive thinkers. Connected brains are creative brains.

Owning a complex, argumentative brain has its advantages. What makes us creative isn't how hard we focus on a task but how well different parts of our brain work together to come up with novel solutions. Transformational creativity, in particular, requires "messy thinking." Coming up with novel ideas—ideas that nobody has seen before—isn't just a simple matter of connecting the dots. Rather, it involves mistakes, ambiguity, and conflict—all resulting from disregarded rules and guidelines.

So, the question remains: Will computers ever embrace such thinking and achieve transformational creativity? I don't know, but we might be better off asking whether anyone would recognize it if one did.

Let's consider the haiku for a moment. Haikus are short, three-line poems that traditionally contain seventeen syllables (though in translated Western form, sometimes even fewer). Haikus date back to ninth-century Japan, where they explored and celebrated prominent religious themes, particularly Buddhism and Taoism. Since then, generations of artists have worked with this art form so extensively that it has transcended Japanese culture, becoming part of world literature.

In short, humans should be pretty good at writing haikus by now. So, here's a test:

> *Early dew*
> *The water contains*
> *Teaspoons of honey*

> *Autumn moonlight—*
> *A worm digs silently*
> *Into the chestnut*

Which one of these haikus was written by a computer, and which one by the seventeenth-century Japanese poet Matsuo Basho—one of the most respected artists of all time?

Hard to say, isn't it? In fact, most people have a hard time answering this question. The first haiku, the one about early dew, was written by Gaiku, a program that starts with "seed words"—that is, opening themes from existing haikus—and then uses complex word-association networks to complete the rest. Gaiku's approach is generally quite effective, though occasionally it does miss the mark. In a recent comparison study, naive subjects were able to distinguish its haikus from human-created ones only 63 percent of the time. That's impressive.

Some creations, however, failed miserably. Take, for example, the following, which one subject argued had to be human-created, because it was "too stupid to be generated by a computer." He was wrong.

> *Holy cow*
> *A carton of milk*
> *Seeking a church*

Let's consider this reasoning again—too stupid to be made by a computer! It highlights exactly what's meant by transformational creativity: the ability to produce a work of art unlike anything previously seen. Would most of us question this "Holy cow" haiku if told by an expert that it's Matsuo Basho's masterpiece, one that changed the way artists look at this special kind of poem? I'm not so sure.

Haikus aren't the only art form being exploited by computers. Programs now write music, draw paintings, and even create Aesop-like fables. For example, a program developed by Paul Hodgson at the University of Sussex improvises jazz in the style of Charlie Parker. Its music is so similar to The Bird's own work that many people can't tell the difference. And a program developed by the architects Hank Koning and Julie Eizenberg uses a representational grammar of Frank Lloyd Wright's architectural style to develop new, never-seen houses that look as though they were designed by the original artist.

Yet, these programs still aren't transformationally creative. They don't break boundaries, and they don't surprise us with unexpected insights into music and architecture. Only their inspirations—Charlie Parker and Frank Lloyd Wright—could do that.

Still, some computer programs are fairly impressive. One humorous example is the program Pato and Perro, which creates cartoons whose two main characters make comments about recent movies. It has elicited more than a few hearty chuckles from this particular author, and it uses input taken exclusively from RottenTomatoes.com. Another good example is artist Harold Cohen's program, which creates pleasing, sometimes unpredictable line drawings. Its work has even been displayed in the Tate Gallery in London, and not just for the sake of novelty.

Each of these arts was once considered too complex for machine intelligence, but every year some new program shows that this is no longer the case. Now the only thing holding us back isn't the size of microchips or the capacity of memory, it's our understanding of what creativity is. What makes an artistic work creative? The answer is subjective, but that doesn't mean the question is unanswerable.

This subjectivity requires that artists—whether made of carbon or silicone—must explain why their work is transformative to convince us that it deserves to be recognized for altering its genre. And that's difficult, not just for computers but for people too. "History is littered with examples of greatness not recognized in its time," says Boden. "Where artists don't accept a new idea, then years later agree that it is valuable. It takes not just familiarity for people to recognize the transformation, but time and comparison. What do musicologists do? Literary critics? [They assess] whether authors like Henry James created valuable work. Or Jane Austen. Both are very different artists, and we assess the value of each for different reasons. But the job of a critic is to recognize the art, and the creativity, and see how each has taken their profession further."

This really gets to the heart of what makes a work transformationally creative, and why original, moving humor is so difficult to come up with. The key is emotional impact. It's the difference between a *ha ha* moment and an *aha* moment (or a *Ha!* moment). The first makes you laugh, the second makes you think of something you hadn't considered before. The reason Gaiku's "Holy cow" haiku falls short of being transformationally creative isn't that it fails to push boundaries. It's that it lacks the intention to be something never seen or heard before. This is a promising idea, one that deserves more attention, but first we need to explore the artist's goal. If the quality of art depends on the intentions of the artist, does absence of intention imply absence of art? Must art break boundaries, or can it simply entertain? In this last section, we'll explore these important questions while also addressing what it means when the quality of art—and of a joke—depends on the goals behind it.

## Keeping Salt Out

All this talk about art might seem a bit heavy for a book about humor. Perhaps this is why most people still view humor the same way the Supreme Court categorizes pornography—we know it when we see it.

It's not easy measuring the worthiness of art *or* jokes. Until computers experience the ambiguity and messy thinking confronted by humans, they won't be able to appreciate the worthiness of any creative enterprise. Recognizing this worthiness is a skill, one that requires that we see how a work of art—or of humor—fits within its larger genre, as well as how the artist struggled to develop it. That last clause is especially important, because computers don't struggle. Their thinking is too linear for that. And that is why they fail.

Mary Lou Maher, former program director of Human Centered Computing at the National Science Foundation, identified three specific components involved in the subjective assessment of creativity. The first is novelty—how different an item is from other members in its class. I love George Carlin's work, and I've probably laughed more at him than any other comic, but when people ask who I like more, Carlin or Lenny Bruce, I always say Bruce. Why? Because Bruce did something that nobody before him had done, or even tried to do. He performed comedy that no one thought possible—or legal. By the time Carlin published his first comedy album, Bruce had already been thrown in jail four times on charges of obscenity. Somebody willing to go through that for his art deserves major points for novelty by me.

The second component is unexpectedness, which is closely associated with surprise. Sarah Silverman is a master of the unexpected. As an attractive Jewish woman, she looks like the kind of comic whose idea of a racy joke involves a priest and a rabbi walking into a bar. Instead, her jokes are some of the crudest ones you'll ever hear. They're racist, and sexist, and blasphemous—and when you watch her you can't help wonder how such language could come from such an ingenuous personality. The epithets she uses are unexpected, even shocking, and they're also the reason her humor is so effective. The contrast between her words and her delivery shows just how stupid and meaningless such epithets really are.

The third component identified by Maher is value, which reflects how appealing an item is in terms of beauty or utility. It's also the most difficult to assess. The first time any of us heard the joke *Why did the baby cross the road? Because he was stapled to the chicken,* it probably scored relatively high on novelty and slightly lower on unexpectedness.

It was an unusual take on road-crossing chicken jokes, making it novel, but we already knew that the ending involves a chicken, diminishing its unexpectedness. The joke's value, however, is as low as you can get, because nobody enjoys this kind of joke anymore, and even fewer enjoy imagining a baby with staple marks.

Not surprisingly, computer-generated humor programs have the most difficulty with value. They lack the real-world knowledge to know what's insightful and poignant, and what's stupid. A lot of people also struggle with this criterion—but that's the point. We struggle because our minds make assumptions, then change them again, then revise them even further. As Richard Wiseman found in his LaughLab competition, the jokes ranked highest in value by some subjects were frequently ranked lowest in value by others. That's because they pushed the subjects' minds to a place where many were uncomfortable going. Good jokes, like progressive art, make us question what we value.

Computers may one day think like people do, and may make discoveries and tell jokes that transform humor, but when they do they won't look like Watson. Instead, they'll have to adopt the same characteristics that allow people to do these things—they'll need to act and think messy. That outcome won't be achieved through simple rules or programs. It will require something entirely different.

I'm referring here to evolutionary algorithms, which rely on the same process that gave us humanity—natural selection. Rather than relying on rules programmed into computer memories, evolutionary algorithms start simply but then modify themselves in small, minor ways. As with natural selection, algorithms that succeed are allowed to survive, and those that fail are replaced in future generations. Computer scientists who use evolutionary algorithms to solve problems don't specify how the solution must take place. Instead, they merely define success. And that depends on what they want their programs to eventually do.

In fact, computers have been using similar unstructured approaches for years, allowing discoveries to be made through unsupervised innovation. A case in point is The Automatic Mathematician, a program that changed the way we look at mathematics more than two decades ago. With an initial database comprising a hundred simple mathematical rules (all more simple even than the rules governing addition and subtraction) and a handful of learning heuristics, The Automatic Mathematician began varying these rules to see what would happen. When variations on the rules worked, they were retained, and those that didn't were discarded. Following this simple process, The Automatic Mathematician re-created a huge library of mathematical rules. For example, without any help it discovered the existence of integers, primes, and square roots. It also discovered Goldbach's conjecture, which states that every even number is the sum of two primes. And then it did something that some would describe as transformationally creative: it discovered a new theorem concerning maximally divisible numbers—numbers unknown even to its programmer.

If a computer can discover a new mathematical theorem starting with only a few basic principles, such as "1 is larger than 0," couldn't a computer evolve to tell a decent joke?

But wait, you might say, even The Automatic Mathematician isn't truly creative. Like Gaiku, it didn't "understand" what it was doing. All it did was produce output—and it did so without any true knowledge of math. This brings us to our final topic, one that points to perhaps the biggest impediment to artificial intelligence research, and to humor development programs too: Can computers actually think?

This may seem a rather philosophical question for a humor book, but it's especially important because, as we've seen, the value of a joke depends on the thinking that went into creating it. The issue isn't whether a computer could have written George Carlin's "Seven Words You Can't Say on Television." It's whether, if the bit *were* written by a computer, would it matter that the computer hadn't been raised in a Catholic home as the son of a troubled marriage, as Carlin was? Would that make the joke less funny?

There's no easy answer to such questions, because by asking if computers will ever be truly humorous, we're really asking if they will ever be conscious and able to appreciate their own funny jokes. That's a tough task, and a pretty good test of conscious awareness too. This issue also raises some deep questions about what it means to "appreciate" a joke. It's easy to assume that one person's phenomenological experience of humor is the same as another's—but there's no proof this is the case. Perhaps joke appreciation is simply a matter of going through the stages of humor processing already described, ending in a resolution that activates some new script or perspective, followed by a final squirt of dopamine to make it all feel good. Could that be all there is to life, the universe, and everything?

Consider a moment this thought experiment created by the philosopher John Searle: suppose you were to program a computer that answers any question it receives in Chinese, so convincingly that anyone interacting with it is certain it knows the language. Does the computer truly know Chinese? Now, suppose you're locked in a room with this computer and are handed slips of paper containing questions written in Chinese symbols. If you used this computer to answer the questions on those sheets of paper, would this mean that you know Chinese too?

Searle's scenario, called the Chinese Room Thought Experiment, is meant to highlight the issue of intentionality—acts of consciousness imparting thought and deliberateness to our actions. According to Searle, you wouldn't know Chinese in this scenario because no actual Chinese thinking is involved. It's an interesting philosophical problem, but one that I have no interest in addressing because I find the whole topic pointless. The real question is whether computers will ever think in the same way people do, and for that I have no answer beyond saying that I don't believe computers must be made of carbon to be creative. Saying that computers must look like us, or think like us, is anthropocentrism. What makes people creative—or sentient—can't just be what we're made of. Rather, it has to be our success at solving problems. This, in my opinion, is interesting enough.

>> <<

I would like to end this chapter with a story from my first year in graduate school, back when I was a young scientist. I was attending a neuroanatomy class taught by Arnold Scheibel, one of the most respected neuroscientists in the country. When someone was needed to examine Einstein's brain to identify the source of his genius, one of the people chosen was Arnold Scheibel. We used to joke that Scheibel invented the neuron, when really he was just among the first to discover how they communicate.

Dr. Scheibel was known for being a rather direct, serious teacher, occasionally making jokes but mostly dispensing vast amounts of information and expecting his students to keep up. One morning, in a complete break of routine, Scheibel opened the class by announcing that at the end of his lecture he would share his recently discovered secret of life. Without a pause he proceeded with his lecture, and for the rest of the session we were left to wonder whether he was serious. Scheibel wasn't one to exaggerate, so his claim seemed real. Finally, with only a few minutes before the class was scheduled to end, he returned to his promise.

"So, now on to the meaning of life," he said, sounding like a computer in a Douglas Adams story.

"The secret is, simply, keeping salt out," he said. Philosophers and religious scholars can question the purpose of our existence all they want, but life serves one main goal, and that's keeping salt on the correct side of our cellular membranes. All neurons are inherently polarized, meaning that they hold a negative charge relative to their surroundings. That charge is maintained by keeping positively charged sodium ions outside the cell body, while potassium and other chemicals are given free passage. When neurons need to communicate, sodium ions are allowed briefly inside so that an electrical current is formed, thus triggering a chemical chain reaction and transfer of information to other cells. If the process breaks down and sodium travels freely through our cellular membranes, our neurons no longer function and we quickly die. That's why without at least some sodium in our diet, we risk severe health consequences, because this transport of salt is essential. Too much sodium is dangerous too, because that threatens the heart, leading to hypertension and even cardiac arrest. Indeed, if there's one thing that life couldn't exist without—or, to put it another way, if there's one thing life was designed to perpetuate—it's the keeping of salt outside our cellular membranes.

"And so, that is the secret of life," Scheibel claimed. The lesson was over. What many neurophysiologists merely call the sodium pump had been elevated to the reason for our existence.

I was fortunate to have the opportunity to share this story once with Margaret Boden, who has written about the sodium pump herself—and she was fascinated, but skeptical too. It seems the solution could just as easily have been ATP, she claimed. Otherwise known as adenosine triphosphate, ATP is an unstable chemical molecule used for energy storage. It's responsible for everything from photosynthesis to biosynthesis and is found in every species regardless of type or complexity, providing a means for organisms to retain energy so that it can be used at future times. "Of course, it's not the perfect example," Boden added, with more than a hint of disappointment. "Because it turns out that very recently they found an organism with constant access to usable energy, I forget where. So, not every organism uses ATP, only 99.99% or something like that. That's the problem with exceptions, because just when you think you have a general rule, some anomaly pops up."

Why did I share this story, especially in a chapter that started out as a discussion of computer-generated humor? Perhaps I wanted to show that the secret of life could just as easily have been a matter of keeping electrons in line, had computers evolved instead of humans. Or maybe I wanted you to question why Scheibel's secret of life wasn't about keeping potassium in.

Actually, I wanted to point out how silly such hypothetical speculation is, in order to appease our computer overlords when they do eventually conquer the earth and enslave us in their sugar mines.

As Ken Jennings said after his loss to Watson: "Karma is a bitch."

# PART THREE

# "So What?"

## BECOMING A MORE JOVIAL PERSON

# 6

## » THE BILL COSBY EFFECT

*The witch doctor succeeds for the same reason
all the rest of us [doctors] succeed. Each patient
carries his own doctor inside him. . . . We are at
best when we give the doctor who resides within
each patient a chance to go to work.*

—Dr. Albert Schweitzer, quoted in
Norman Cousins's *Anatomy of an Illness*

Now it's time for another shift in topic. The first three chapters addressed the *What is?* question of humor: What makes us laugh, and how do our brains turn conflict into pleasure? The two chapters after that addressed the *What for?* question: What purpose does humor serve, and what does it say about who we really are? Both of these sections provided important background for understanding why we laugh, but there's an even more important question we haven't addressed yet. I call it the *So what?* question: Why should we care what humor is, and how does it influence our physical, psychological, and social well-being?

Studies show that humor improves our health, helps us get along better with others, and even makes us smarter. In the next three chapters we'll see how. Along the way we'll attend a comedy show, watch a corporate wrestling event, and see how listening to Bill Cosby raises our threshold for pain. And it all starts with a man named Norman Cousins, who was told by his doctors that he had a 1 in 500 chance of surviving a debilitating disease and ended up beating the odds through comedy. In fact, he laughed himself out of an illness.

## NORMAN COUSINS

Cousins's story begins in July 1964 at a political conference in Moscow, where as chairman of the American Delegation he was charged with attending formal meetings on improving cultural exchange between the Soviet Republic and the United States. This involved many long evenings at social events and formal dinners—a stressful schedule considering that they were held in a country where he didn't speak the language. He was also exposed to an unhealthy atmosphere—literally. Mid-twentieth-century Moscow was notorious for its dirty air and water, and Cousins's hotel room was located in the center of town, right next to a housing construction site. Diesel trucks spewing fumes twenty-four hours a day left him feeling nauseated each morning. By the time he returned to America, his joints ached. Pretty soon he couldn't move his neck, arms, or legs. His body had been overcome by a debilitating malaise.

Growing concerned, Cousins finally saw a doctor and was told that he had contracted a severe collagen illness called ankylosing spondylitis. Collagen is the fibrous substance that binds our cells together, and Cousins's was disintegrating. Without it, he would become unable to move.

"In a sense, then," recounts Cousins, "I was becoming unstuck."

His outlook was dire. Specialists told him that his only hope was to fight the pain using drugs, but Cousins knew that when drugs become the focus of a treatment, that's a problem. "People tend to regard drugs as though they were automobiles," he complained. "Each year has to have its new models, and the more powerful the better."

Another troubling aspect of his treatment was the disruptive way the medical staff addressed his illness. Once, four separate technicians took large vials of his blood in a single day. Taking so much blood—even from people who are well—usually isn't a good idea, and Cousins wondered if maybe the treatment wasn't doing more harm than good. He was fed mostly processed meals rather than a healthy balance of natural food. His sleep was frequently interrupted for tests that could easily have waited until the morning.

It was about this time that Cousins decided that, instead of trusting the doctors, he would laugh.

First, he left the depressing surroundings of the hospital and checked into a hotel, which was not only more cheerful but cost only a third as much. Then he got to thinking: What could he do to help himself? Since traditional medicine wasn't going to cure him, what other approaches could he take? It was then that Cousins began to consider the effect of stress on medical recovery. Stress likely contributed to his illness, as well as hindered his treatment, so it seemed reasonable to question whether the effect worked in both directions. "If negative emotions produce negative chemical changes in the body, wouldn't the positive emotions produce positive chemical changes?" he asked. "Is it possible that love, hope, faith, laughter, confidence, and the will to live have therapeutic value? Do chemical changes occur only on the downside?"

One way to find out was to put himself in a good mood—and to do that, Cousins began a systematic plan for laughing. He started with films of the old practical-joke program *Candid Camera* (similar to *Punk'd*, but without Ashton Kutcher). This wasn't easy, since DVDs and Blu-rays hadn't yet been invented and the only way to watch these shows was to use a motion-picture projector. But he was able to borrow one from a friend, along with several Marx Brothers films, and anything that made him laugh became part of his treatment.

Cousins watched the films regularly, every day, and despite being in pain he discovered that he was still able to laugh. Not only that, but the laughter was more effective in combating the pain than aspirin or any of his other analgesics. "Ten minutes of genuine belly laughter . . . would give me at least two hours of pain-free sleep," he wrote.

Amazingly, after a little over a week of rest and laughing, Cousins was able to move his thumbs again, something his doctors previously thought impossible. After several months, he could grab books from atop bookshelves, and as more time passed he even hit tennis balls and played some golf. His disease hadn't disappeared—one shoulder and both knees were still causing him occasional trouble—but given his initial prognosis, his recovery was incredible. Cousins went on to live twenty-six more years.

Cousins's recovery was uplifting and positive, yet also rather troubling. His rejection of doctors, hospitals, and the newest drugs in favor of a more holistic approach probably saved his life. But if you were in the same situation, would you have had the strength to make the same choice?

We've all seen medical "cures" that are far from scientific. However, while it's easy to blame modern medicine for being impersonal, it's wrong to think of doctors as closed-minded people. It's the rare doctor who wouldn't do anything humanly possible to help a patient. Alternative-medicine approaches, such as laughing, are alternative for a reason—they have yet to be proven as beneficial. They haven't been ignored. Quite the opposite, as we'll see in this chapter, laughter as a medical treatment has been studied quite extensively. Doctors just don't prescribe it to their patients for the same reason they don't recommend other alternative medicines, such as acupuncture or large doses of vitamin C. Research results have been mixed.

This chapter takes a holistic view of humor and its effects on the human body. So far we've seen that our brains use conflict like our muscles use oxygen, or cars use gasoline. Humor empowers us to make decisions and take pleasure in a complex world. But the benefits don't end there. Humor is also a form of exercise, keeping our minds healthy the same way that physical exertion helps our bodies. But like jogging in a smog-filled tunnel, humor used improperly can do more harm than good.

## THE DOCTOR INSIDE

"Each patient carries his own doctor inside him."

This is the claim that best describes Cousins's philosophy, one taken directly from his book, *Anatomy of an Illness*. In his recounting, Cousins describes both his laughing treatment and the responses he received from doctors, relatives, and friends after they heard news of his recovery. Everybody seemed to have an opinion about what had caused his ankylosing spondylitis to remit. Some thought Cousins had simply willed the disease away through positive thinking. Others argued that his recovery was an anomaly—one that randomly occurs in only one in a million patients and shouldn't be interpreted as a blueprint for future success. Still others simply congratulated Cousins for having the courage to control his own medical destiny.

At one point in his book Cousins equates laughter with internal jogging—a great analogy. We know that laughter benefits the body because it's an aerobic exercise. Highly controlled measurements have shown that laughter expends between 40 and 170 kilocalories per hour. Lots of work had equated that to other forms of exercise, with the most common being that a hundred laughs are roughly equivalent to ten to fifteen minutes on an exercise bike. How you measure number of laughs I have no idea, but it still sounds like a good time to me.

One way jogging improves our health is by pushing our hearts to work harder, and laughter relies on the same mechanism. As researchers have shown, both systolic and diastolic blood pressure rise during exercise. The same thing happens during laughter. Sometimes these changes last no longer than a single heart beat, and sometimes they last much longer, but that elevation is critical because the more we work our heart, the lower our resting blood pressure—and the less our hearts have to work the rest of the time.

This benefit can be long lasting, too. Two measures of blood flow—carotid arterial compliance and brachial artery flow mediated dilation—can remain elevated for up to twenty-four hours after viewing a laughter-inducing comedy.

One scientist who knows a lot about the benefits of laughter is Michael Miller from the University of Maryland. His specialty is vasodilation, which refers to the widening of blood vessels. The reason elevated blood pressure during exercise is healthy is that it helps our blood vessels stay flexible. Healthy vessels relax or constrict depending on our level of activity, but unhealthy ones remain stiff and tight, restricting blood flow at the times when we most need it.

Two of the greatest threats to our health are vasoconstriction and reduced vasoreactivity. Often caused by stress, these conditions lead to narrower blood vessels, reduced blood flow, and a decreased ability to vary the amount of blood delivered throughout the body. In many people they also lead to coronary disease and stroke. Doctors recommend frequent exercise because aerobic activity loosens up the blood vessels, making them more pliable. And so does laughter, according to Michael Miller, who presented his findings on vasoreactivity to the American College of Cardiology in 2005. Specifically, laughter decreases stiffness and increases vascular reactivity, thereby increasing blood flow to the areas of the body that need it.

Miller's study examined twenty men and women, all with roughly the same level of heart health when the study began. Subjects watched the opening scene of either a stressful movie such as *Saving Private Ryan* or a comedy such as *Kingpin*. Before and after each movie, measures of vasodilation were taken, assessed by tightening and then releasing a blood pressure cuff on the arm. By aiming ultrasound devices at subjects' arteries, Miller also measured how well the arteries "rebounded" after the cuff's constriction, thereby letting him know how flexible or unresponsive they were. He expected the stressful movie to lead to less responsive blood flow, as had been seen in numerous previous studies. The question was whether humor would have the opposite effect, and low long it might last.

Miller found that fourteen of the twenty subjects experienced reduced artery size following the stressful viewing, leading to reduced blood flow. Even more impressive, however, was the change seen in the humor-watching group. Among those subjects, all but one showed *increased* artery size, with an improvement in blood flow of more than 20 percent. This change lasted long after the movies concluded.

It has always fascinated me that our bodies don't get stronger through rest but by exertion. Muscles increase in mass first by breaking down, then rebuilding. Blood flow is improved when blood pressure is first increased through exercise, then allowed to return to a lower state, the blood vessels more relaxed than if they hadn't been worked. Laughter has the same effect, and that makes us stronger and more capable of dealing with challenges later on.

Our hearts aren't the only things benefited by laughter, either. Research shows that laughter suppresses glucose levels in diabetics, helping to prevent diabetic neuropathy. It also improves immune system function, reduces chemicals associated with joint swelling in arthritis patients, and even helps allergy sufferers combat dermatitis. In short, the mirth associated with laughter leads to positive physiological changes throughout the body.

A big challenge for psychologists and doctors is identifying how cognitive states, like mirth, lead to physical changes in the body. We know, for instance, that exercise puts us in a good mood because it leads to the release of dopamine, which gives us pleasure. That's a physical act leading to psychological change. But does cause and effect work in both directions? Can improved state of mind lead to physiological change?

Fortunately, it can. Consider, for example, immunoglobulin A. This antibody is one of our immune system's first lines of defense against invading organisms such as bacteria, viruses, and even cancer cells. Though our bodies produce several different kinds of such antibodies, they all work the same way—first, by identifying and targeting the foreign body and, then, by either neutralizing it or tagging it for attack by other defense mechanisms. Studies show that watching funny movies and listening to stand-up comedy significantly increase immunoglobulatory response—and so does being in a humorous mood. A similar effect has also been associated with natural killer cells, which, in addition to having a very impressive name, help stave off diseases like cancer and HIV. Watching movies like *Bill Cosby: Himself* or *Robin Williams, Live at the Met* has been shown to increase killer cell counts by up to 60 percent.

By now, you might think that laughter is the best thing you could do for your body. It improves cardiovascular health, boosts immune response, and even activates cells that attack invaders like teams of highly trained Navy Seals. For Norman Cousins, it helped overcome the rheumatic disease ankylosing spondylitis. If humor does all these things, what's to keep us from laughing our way to immortality?

It's a good question. And the answer is—humor doesn't actually help us live any longer. In fact, it does just the reverse.

I was surprised when I first learned this interesting fact, and you probably are too, given the findings discussed above. Laughter certainly has its benefits, but these don't necessarily ensure a longer life. It's a disappointing fact, but it's important to recognize because laughing isn't a panacea. It's an activity much like jogging or jumping rope. Used responsibly in the right circumstances, it can be a great protector. But if used without good judgment it can be as dangerous as running a marathon barefoot.

Let's look at two studies to see why. The first involved nearly the entire population of the Norwegian city of Nord-Trøndelag—more than sixty-five thousand people in all—agreeing to take three tests in the name of science. One measured their sense of humor, with questions like "Do you consider yourself to be a mirthful person?" A second test assessed their bodily complaints. Essentially a psychological survey of health, it asked respondents questions about common complaints such as heartburn, nausea, even constipation, rating the frequency of each on a three-point scale. The third test involved a blood pressure reading and a measurement of body mass index.

As expected, the researchers found a significant relationship between overall health satisfaction and sense of humor. It wasn't huge, only a correlation of 0.12, but it was positive, suggesting that the more subjects appreciated humor, the happier they were about their health. However, sense of humor had no benefit on actual measures of health such as blood pressure or weight. Though older individuals tended to have higher blood pressure and reduced sense of humor, when the factor of age was controlled, humor showed no relationship with heart health.

That's a disappointing finding, of course, but there's a lot more to health than blood pressure. Perhaps the important question isn't how much humor helps our bodies but how much longer it allows us to live. Longevity is really the best measure of health, so shouldn't we measure that?

I believe we should, and that's why I also present this study by psychologist Howard Friedman of the University of California at Riverside. It took over sixty years to conduct because Friedman didn't just give people surveys or blood pressure readings. Instead he enlisted more than fifteen hundred children, all roughly eleven years of age when the study began in 1921, and waited for them to die. Actually, that's a rather morbid description, because his true goal was to follow these middle-class schoolchildren over the course of their lives to monitor changes in their health. He wanted to see if there were any connections between their personality traits and their longevity, and also if this longitudinal approach could uncover relationships that weren't observable using simple physiological measures. Fortunately, though some children disappeared over the years, more than 90 percent of his sample checked in with him regularly throughout the years.

After revisiting his subjects in 1986, over sixty years after his study began, Friedman concluded that one personality trait significantly predicted longevity—conscientiousness. This trait reflects how prudent and thoughtful a person is when dealing with others, and it appeared to raise survival rates by as much as five years. Self-confidence didn't show this effect, and neither did sociability. In fact, the only other factor affecting how long Friedman's subjects lived was sense of humor, which apparently didn't prolong lives but shortened them.

The effect wasn't as strong as that of conscientiousness, but it was still impressive—the humorous people in Friedman's sample lived shorter lives than everybody else.

The reason for this inverse relationship is hard to explain, though it may have to do with the possibility that humorous people don't take better care of their bodies. For example, other studies show that, compared to nonhumorous people, they smoke more, put on more weight, and have a greater risk of cardiovascular disease. And then there's the issue of neuroticism, a personality trait often seen among humorous individuals (as we saw in Chapter 4). Neuroticism has been shown to increase risk of mortality by up to 30 percent. So, even though Cousins's strategy bought him an extra twenty-six years of life, it doesn't seem to work for everybody.

But don't give up hope yet. My goal is to show that humor is an essential protective mechanism. It may not promise longer lives, but it does enhance psychological health and shield us from pain. Which brings us to something I like to call *the Bill Cosby Effect*.

## THE BILL COSBY EFFECT

Imagine that you are about to have orthopedic surgery to correct a sprained disk in your back, when you are approached by someone claiming to be a graduate student majoring in communications arts. He tells you that he is doing his thesis on the effects of various media on pain, and asks if you would be willing to participate in his study. If so, then after granting him access to your medical records you will be shown movies to see how they affect your postsurgical recovery. It sounds like a fair deal, and you agree.

Fortunately your surgery is a success, and when it is over you are wheeled into a recovery room where you see the experimenter again. This time he is accompanied by a VCR player. He gives you a choice of humorous movies to watch, some new and some old, and on the list you see several favorites—*Weekend at Bernie's, Naked Gun,* and *The Odd Couple.* In the end you choose *Bill Cosby: 49,* a concert film of Cosby's stand-up in which he talks about topics like aging, wives, and children. That evening you're presented with the same selection, except this time you choose *A Fish Called Wanda.*

The entire process is repeated the second day, and soon you're feeling pretty good about your decision to participate. The experimenter asks a few awkward questions, particularly about the amount of pain you experienced after surgery and how you feel about your recovery, but these intrusions are minor. By the third day you've healed well enough for you to rehabilitate at home, so the hospital releases you to outpatient care. The experiment is over, and you thank the researcher before leaving.

Your own participation in this experiment was imaginary, but in real life this was one of the most important humor studies of all time. Conducted by James Rotton, a psychologist at Florida International University, it examined the effect of humorous movies on pain tolerance. There were seventy-eight participants in all, and the luckiest ones were those who watched funny movies. Other subjects saw movies with less emotional impact, such as *Dr. No* and *Labyrinth.*

I say "luckiest" because the subjects who viewed funny movies experienced less self-reported pain than those who watched dramas. Specifically, they felt better about their postsurgical condition and requested 25 percent less medication than their peers. By the second day, this translated to a third less pain, even with the reduced drugs.

This finding is especially important because it shows how humor doesn't just make us healthier—it improves our quality of life. And it's all the more meaningful given that Rotton's design used actual patients with actual pain, a rare occurrence in the world of science. Conducting research in real-world settings is always preferred, but it's a challenge because universities are notoriously sensitive about protecting research subjects. It's the rare institution that allows its scientists to "hurt" its subjects, making pain a difficult topic to study.

This isn't to say that pain tolerance can't be studied in laboratories too. A benefit of laboratory studies is that they allow for complete control over the procedure, something Rotton didn't have. He had no way of manipulating the actual pain his subjects endured because he didn't perform any actual surgeries or cause the injuries requiring treatment. Experimental pain studies, on the other hand, allow for such variables to be controlled and ensure that all subjects feel the same level of discomfort to begin with. One way to accomplish these goals is to use a cold pressor test.

The cold pressor test involves inserting your hand in a bucket of ice water chilled to 35 degrees Fahrenheit, just above freezing. Though cold and painful, this stimulus doesn't do any actual damage or cause frostbite, it only makes you really want to remove your hand. The amount of time you keep your hand inside the bucket becomes your personal measure of pain tolerance, and by administering the test multiple times—for example, before and after some experimental treatment—scientists are able to look at how your pain threshold changes over time.

Studies using the cold pressor test confirm that watching comedies does indeed give us protection from pain. This is what I mean by the *Bill Cosby Effect*—simply viewing a recording of humorous stand-up can increase the amount of time we're able to keep our hand in ice water, from 36 to 100 seconds.

Norman Cousins claimed that laughter is a natural pain killer, a better treatment than any manufactured drug. That would be a nice story, and a satisfying one too, but life is more complicated than that. Indeed, as it turns out, watching gruesome horror movies, filled with blood and gore, has the same effect. People who watch such movies are able to hold their hands in ice water for nearly two minutes, an increase of more than 300 percent compared to baseline. So the analgesic effect of movies isn't limited to humor.

How could that be? This whole time we've been exploring the benefits of positive mood, and now we see that fear has the same effect? It does, and to explain why we need to revisit the Rotton pain study. From his original finding you might have guessed that watching a funny movie is like taking aspirin—or, in the case of exceptionally funny movies, codeine. But note what happened when the subjects in that study were shown movies other than the ones they selected.

One thing I didn't share earlier about Rotton's study is that he asked all his subjects what kind of movie they preferred, but only half of them actually viewed the movies of their choice. The other half—those in the "no choice" group—were shown movies that they *hadn't* selected. Interestingly, he found that for the latter subjects, watching funny movies didn't decrease requests for pain medication. Instead it increased them. In fact, the "no choice" subjects asked for more than twice as much pain medication as did those whose choices were honored. So, watching funny movies wasn't enough to increase pain tolerance. Subjects also had to watch the kind of humor they liked, and had to feel as though they were in control of their own moods.

In short, laughter—or, more specifically, positive affect—isn't what confers benefits. What matters is our emotional engagement. The subjects in Rotton's study who watched movies they didn't choose probably didn't find the films funny because they weren't emotionally invested in them. Our minds need emotional engagement just like they need exercise. Without that engagement, we become passive to our environment. And a passive mind is an unhealthy mind.

The reason both comedies and tragedies lead to greater pain tolerance is that our minds are exercised by each. When we laugh, just as when we cry, our bodies experience emotional arousal. This effect is both engaging and distracting, strengthening our bodies—and our minds—for what is to come, much like a boxer lifts weights before a bout. Earlier we explored the question of how psychological states induce physical changes in the body—and now we see that, by providing our brains with exercise, humor prepares them for greater stressors down the road. Conflict can be a good thing, so long as it's harnessed properly.

This emotional-exercise view of humor also explains why watching comedies is sometimes more beneficial than meditating or listening to calming music. For example, simply viewing an episode of the sitcom

*Friends* has been shown to reduce anxiety three times more effectively than sitting and resting. Our brains want to relax and overcome stress, but they need to stay active too. Simply allowing them to be dormant does no good.

Like physical exercise, humor takes many different forms, and not all of them are created equal. Humor keeps our brains and bodies active, but not all activity—whether physical or mental—is beneficial. Earlier we noted that there are no fewer than forty-four separate types of humor—wit, irony, and slapstick being just a few. But humor styles vary too, based on psychological motivations rather than punch lines. One example is affiliative humor. People with high degrees of affiliative humor enjoy saying funny things, sharing witty banter to amuse friends, and joking to reduce interpersonal tension. As you might guess, affiliative humor is considered a positive humor style, meaning that it leads to constructive psychological and social behaviors.

Another positive style is self-enhancing humor, which characterizes people who enjoy funny outlooks and who laugh in order to see the bright side of troubling situations—anything to keep a positive attitude. People who score high on measures of self-enhancing humor tend to have high self-esteem and to be conscientious. And as we've seen, that last personality trait is especially important for health because it's the only personality factor that predicts longevity. So humor can indeed help us live longer, as long as it's the right type.

The idea of humor styles was developed by Willibald Ruch as a way of characterizing humor in everyday life. They're assessed using a test called the Humor Styles Questionnaire, which identifies two negative styles too. First there's aggressive humor, which involves sarcasm, teasing, or ridicule. People who rely on aggressive humor try to build themselves up at others' expense, and not surprisingly also score high on tests of hostility or aggression. And then there's self-defeating humor. To understand this style, you need only think of Rodney Dangerfield. Rather than putting down others, self-defeating humorists target themselves, often as a defense mechanism for low self-worth.

Whereas positive humor styles increase feelings of self-worth and conscientiousness, and possibly even improve longevity, negative humor styles have the opposite effect. People who use self-defeating humor tend to experience depression, anxiety, and low self-esteem, and those who use aggressive humor often adopt poor coping mechanisms. It's easy to see how this might have adverse long-term effects on longevity.

In short, humor can either improve or harm our health, depending on how it's used. Dealing with conflict in positive ways, such as laughing to put ourselves in a good mood, is probably as important as getting on that Stairmaster three times a week. Laughing negatively at ourselves or taking a dark, sardonic attitude—well, you might as well start drinking and smoking too.

In my life I've run two marathons, once when I turned twenty and the other when I turned forty. For both races I was unprepared but I ran them anyway, the first time to impress a girl and the second time to trick myself into believing I wasn't middle-aged, because that same girl—now my wife—wouldn't let me buy a Porsche. The funny thing about both experiences is that I just missed my goal each time (under four hours), mostly because I hadn't trained well enough. But even so, after the race was over I didn't increase my training. Why? Because the race was over. Exercise improves our fitness only *after* we've finished training.

If laughter is like exercise for the mind, then we should expect mental training to work the same way. And it does, as we'll see in what I like to call the *Faces of Death* test. Arnie Cann is a social psychologist from the University of North Carolina, where he studies how people recover from traumatic experiences. Though we humans are remarkably resilient, sometimes that resilience breaks down. When that happens we get sick or feel depressed—or, in extreme cases, we suffer conditions like posttraumatic stress disorder. One way to protect ourselves from stress is through humor. To explore this possibility, Cann performed an experiment using the most gruesome film in the world, one that prides itself on being banned in over forty countries.

My guess is that if you're over the age of thirty, you've heard of *Faces of Death*. It was released in 1978, and nobody had seen anything like it before. It showed scene after scene of horrific deaths, all presented as actual footage with Rod Sterling-like narration. Men are set on fire, families are doused with napalm, bicyclists are run over by tractor trailers—all in Technicolor splendor. Though the advent of YouTube and allegations of fakery have diminished the movie's impact over the years, it's still quite gory. Few people are able to watch it for more than a few minutes without feeling ill.

Which is why I find it so impressive that Cann was able to convince his university review board to allow him to show subjects a twenty-minute excerpt of the movie, just to see what would happen. Some subjects started with sixteen minutes of stand-up comedy and then viewed scenes of death afterward. For them the humor was intended to provide protection, a sort of inoculation for the gruesome scenes to follow. Others viewed the comedy last. For them it was a form of a recovery, a chance to allow their minds to return to normalcy. A third group of subjects simply watched a travel documentary. Sometimes the documentary came before the scenes of death, sometimes after.

Cann's results showed that the stand-up helped subjects deal with the stressful movie, specifically by lowering the perceived tension. However, those benefits were limited to one group in particular—namely, the subjects who saw the comedy first. In fact, the study showed almost no benefit from *ending* with a comedy because by then it was too late. The only benefits were seen when the subjects were put in a good mood as soon as the experiment started.

This last finding is important because it shows that humor isn't so much a magic cure as a form of prevention. Like exercise before a race, it prepares our bodies for the stress yet to come. This may seem counter to Cousins's experience, since he didn't start laughing until after he was diagnosed. But bear in mind that except for providing some mild exercise, laughter doesn't change the body directly. It works through the mind. It creates a protective outlook, and this is what bolsters our immune system and helps us move past gruesome death scenes. This outlook is key, helping to put Cann's subjects in a good mood to prepare them for the gore that followed. It's also what Cousins cultivated by secluding himself in his hotel room, refusing to accept the odds his doctors gave him.

A neuroscientist friend of mine once shared with me what he called evolution's greatest gift to living creatures: the ability of the brain, at times of extreme damage to our bodies, to stop releasing chemicals associated with pain and alarm. Instead, it releases endorphins, nature's equivalent of morphine. From an evolutionary standpoint it's hard to see the benefits of such a release. It doesn't increase energy, and it doesn't promote recovery either. It simply provides psychological comfort, making our most trying moments peaceful instead of terrifying. Apparently nature thought that was important enough.

I share this story to introduce the concept of positive outlook, which is also related to humor. I'm not talking about enduring pain associated with bear or tiger attacks; rather, my point is that humor helps us overcome psychological injuries. Humorous people experience the same number of stressful events as everybody else, and we know this because scientists have actually counted. However, as researchers have shown, people who are quick to laugh tend to forget those stressful experiences more quickly than those around them. Humor also helps us ignore the events in our lives that might otherwise cause us pain or harm. Humorous people may not experience easier lives than those around them, but they often feel as though they do. They're able to block out negative experiences when they're over and to move on.

Perhaps this is why doctors are finally taking Cousins's advice and incorporating humor as part of their medical treatments. Comedy Carts are springing up in hospitals across the country. For example, the one at the St. Jude Children's Research Hospital in Memphis, Tennessee, which distributes whoopee cushions and funny movies to keep patients laughing. Or the Therapeutic Humor Program at Rochester General Hospital in New York. It distributes comic books and videos to its patients, who report a 50 percent decrease in stress as a result of the program. And then there's the Big Apple Circus, which sends "Clown Care Units" to New York City hospitals to visit sick children and their families.

When the movie *Patch Adams* was released in 1998, several reviewers complained that it downplayed the importance of medicine in recovery. Laughter and positive attitude are certainly beneficial, they said, but they're of little help if the patient dies in the end. That's true, but it's also true that patients receiving the best drugs and medical treatments sometimes die despite the best efforts of doctors and hospitals. Humor alone may not keep us healthy, but it can reduce the amount of pain in our lives—whether real or perceived. It can also strengthen our heart and immune system and, assuming that we use it positively, improve our psychological well-being too. So laughter really *is* the best medicine, so long as it's mixed with exercise, a healthy diet, and an occasional dose of penicillin.

Humor is a lot like changing a baby's diaper—it doesn't necessarily solve all our problems, but it sure does make things more pleasant for a while.

# » Humor Dances

*I've discovered that getting a laugh is more a trick of timing than of true wit.*

—Gore Vidal

Now that we're over forty, my wife Laura and I don't spend many late evenings out anymore. The transition was slow but unmistakable; when we were young we frequently went out to dinners, movies, and comedy shows. Then, as jobs and other responsibilities started competing for time, these activities began to change. We became more inclined to get up early for hikes or bicycle rides. Late-night trips to bars turned into quiet evenings with close friends. We still went out to dinners and clubs, but they were no longer the kind with live bands. They were ones we'd been to before, where we already knew the menu.

So, when I asked Laura if she'd like to go with me to a comedy club as research for this book, I expected her to balk. Boy, was I wrong.

"Let's go tonight," she recommended. She didn't ask who the headliner was or how late the show would be starting. "What should I wear?

Should we get dinner beforehand? Do you think any of our friends would like to join us?" I answered each question as best I could, trying not to overthink whether our transition to adulthood had been a mutual decision. She was in as soon as the offer left my mouth.

We ended up going to Magooby's Joke House, which is one of the most popular comedy clubs in Baltimore and only a short distance from our home in eastern Maryland. Located in a four-hundred-seat theater with tall ceilings and stadium seating, it was the kind of place where everybody has a great view. Like most clubs, it offered a variety of cheese- and potato-related appetizers as well as an elaborate menu of drinks with names like "The Blind Pirate" and "Screwy Light Bulb." So the night was off to a good start.

The first performer was the club's emcee, Mike, and though he wasn't bad he didn't make me laugh much. His comedy was different from the kind I usually enjoy watching, with jokes that were safe crowd-pleasers and a style of delivery using exaggerated movements and facial expressions. At one point, Mike quipped about Facebook and how stupid he thought it is. Then, close to the end of his act, he got several cheers by asking whether anybody in the audience hated the Pittsburgh Steelers. In Baltimore, that's like asking a crowd of Ohio State undergrads if anybody hates the University of Michigan.

The next act was a little better, but already I was looking forward to the headliner, Rich Voss. I knew of Voss because I had watched *Last Comic Standing,* an NBC show where professional comics performed each week and got voted out one by one, *Survivor* style. Ratings were bad and the show was eventually canceled, though I remember Voss because he was one of the funniest and most likable contestants from the opening season. A regular bit involved the comedians giving recorded interviews, and Voss always gave his in a bathtub, usually accompanied by one of the other comics, David Mordal. It was absurd and awkward and, to me, utterly hilarious. So I was excited to be at Magooby's that night.

But as soon as Voss began his act, I could tell Laura was not impressed. He began with a few racial jokes and insults thrown out to the audience, a confrontational style that I suspect was intended to get people worked up early. Then he moved to safer topics like teenage daughters and spousal flatulence, and although I laughed frequently, Laura barely snickered. She was still having a good time, but clearly she wasn't connecting with Voss's dark comedic style. At one point he made a comment about dating, and how men don't just date a woman but all of her friends, relatives, previous boyfriends, and every expectation she has had since childhood. I knew Laura must have found this joke funny because it was exactly her type of humor, yet it got only a chuckle, nothing more. Now it had become personal. She clearly just didn't like Voss.

After the performance, we agreed it had been a great evening and that we should do it again soon. Then Laura remarked that she would return simply to see the emcee, the performer I had disliked. He was hilarious, she said. When I remarked that I didn't think his jokes were original, she said she didn't either, but that she loved the way he delivered them. It was as if we had experienced completely different acts.

It's not that Laura was offended by Voss's humor. Though his jokes were frequently off-color, my wife—a retired military officer who has spent significant time at sea in the North Pacific—isn't easily offended. In short, there isn't much she hasn't seen or heard. And though I have explicit instructions not to make her sound like a longshoreman, let's just say that when Laura is prodded, her profanity could make Mark Twain blush. So what gives?

Clearly, each of the performers deserved to be onstage and each got plenty of laughs. The difference was really one of connectedness. I never connected with Mike the emcee and didn't like how he manipulated his audience. Laura never connected with Voss and didn't enjoy his stern, New Jersey style. These differences highlight the social nature of humor and how much it involves relationships between people.

This chapter explores those relationships. Earlier in the book we looked at how humor violates expectations at a psychological level, leading to revised scripts. Now we'll see how expectations exist on a social level, too. Successful comics manipulate their audiences by controlling their expectations, which for Mike the emcee meant starting with a few simple crowd-pleasers, and for Voss meant insulting various members of the audience. Each approach shaped how the crowd would react to the performers' quips and punch lines. These styles allowed relationships to form—but in each case, damaged connections meant failed humor. Over the next several pages we'll explore why, showing how humor takes advantage of the most challenging aspects of our social relationships, such as subtlety, ambiguity, and conflict. We'll also see how humor brings us together by cultivating shared expectations—and then destroying them.

## HUMOR AND DANCING

The psychologist and philosopher William James once said that common sense and humor are the same thing moving at different speeds. Common sense walks, but humor dances.

Dancing is indeed the perfect analogy for how humor works. Humor, like dancing, is by nature a social phenomenon. Try telling a funny story in an empty room and you'll see what I mean. Without having other faces to look at for a reaction, you'll find that the joke isn't a joke at all. Humor requires both a teller and a receiver, and its success depends on how well one influences the thoughts and expectations of the other.

The dancing analogy also highlights the important but elusive role of tempo. With dancing, there's always a clear beat. Humor has a beat too, and we call it comedic timing, but there's no rhythm section, only our instincts and our ability to read the audience. So, uncoordinated guys like me, who may not be able to dance but can fake our way through a song by biting our lip and listening closely for the bass line, have no safety net. Put us in front of an audience and ask us to tell a joke, and we might as well perform a mamba wearing earplugs.

It's worth addressing comedic timing because, as noted earlier, humor requires a connection between people, and that means being in sync with our audience. Some comedians, like Robin Williams, tend to speed up before a punch line. Others, like Steven Wright, slow down.

Perhaps that's why humorologists liken comedic timing to jazz. Improvisation is key for both, with the onset and duration of each note depending on all the ones that came before it. This makes for playfulness and the constant risk of surprise.

One of the few actual experiments to measure comedic timing was conducted by our old friend Salvatore Attardo, who, as we learned in Chapter 2, also developed the General Theory of Verbal Humor. Attardo recorded ten speakers as they performed jokes, and then he broke down those recordings in three important ways. First, he measured their rate of speech—essentially how fast the joke-tellers talked during the setup and the punch line. Second, he measured their pitch and volume, looking for changes that signaled upcoming humorous turns. Third, he looked for pauses, ranging from less than a fifth of a second (200 milliseconds) to more than four times that long. His hypothesis was that speakers would pause before delivering punch lines and that the punch lines themselves would be delivered faster and louder than the rest of the joke.

Sadly, his hypotheses weren't borne out. Not a single aspect of the punch lines was different than the rest of the joke.

"It caught me absolutely by surprise," says Attardo. "It was to be one of those studies where you expect to confirm what everybody already knows is true, then everybody applauds and moves on. But we found just the opposite. . . . It took quite a bit of work to show that we hadn't screwed anything up, though now it's pretty well recognized that these markers don't distinguish punch lines like we thought they did."

This finding countered what scientists call the folk-theory of joke delivery. Folk-theories are beliefs that everyone "knows are true" without ever having seen the proof. "We use only 10 percent of our brains." "Blind people hear better than those with sight." "Subliminal messages influence our behavior." The problem with such beliefs is that even though they're common, they're also incomplete or wrong. Yes, a small percentage of the brain possesses functions predetermined at birth, but that doesn't mean the rest isn't important too. Blind people do sometimes experience superior hearing, but only when the blindness occurs during infancy, while their brains are still plastic. And unless you're a character in a bad sitcom, being exposed to a subliminal message won't make you do something you wouldn't ordinarily do, like tell off your boss or make a public scene. It might make a word come to mind slightly faster than usual, but there's little proof that the effects are broader than that.

The folk-theory of humor is that punch lines are distinguished by pauses and higher pitch—a belief that, like those other folk-theories, contains a small kernel of truth. The complication comes in the form of "paratones," which is what linguists call spoken paragraphs. Paratones tend to end with *lower* volume and pitch, not higher like jokes. Since jokes often come at the end of paratones, any changes in volume or pitch cancel each other out. This explains why Attardo saw no differences in his measures: the jokes weren't salient enough to overcome the speakers' natural tendency to end on a low note.

Although timing effects don't show up experimentally, we still know they exist. We've all heard people butcher a joke by skipping a needed pause or speeding up just when they needed to slow down. Those pauses and changes in tempo convey important information. Indeed, effective communication involves a lot more than just words; it also depends on what's left unsaid, or what's implied through hesitation and changes in pitch. These cues introduce many layers of meaning, and as we've seen, humor is all about multiple meanings. Expert comedians use pauses and tempo changes to build up expectations and signal upcoming turns, and without those manipulations there's no humor. Just the prolonged telling of a story.

Which is why we wouldn't expect to measure the funniness of a joke using pauses and inflection, because these are only symptoms of a much broader phenomenon. That phenomenon is ambiguity, which occurs not just within our brains but between people too. Punch lines aren't the bang of the jokes, only the tools we use to bring on a final resolution. The build-up—filled with pauses, changes in volume, and all sorts of other subtle indicators—is where the humor starts, because this is where the ambiguity is sowed. You can't just look at a specific part of a joke to find the humor, because it's everywhere.

To see how humor depends on more than just the punch line, just listen to any funny joke or story and note when people laugh. Laughter almost always occurs over the course of entire jokes, not just at the end. A recent survey of nearly two hundred narrative jokes, which are typically lengthy and told in natural conversation, showed that the majority involved setups that elicited laughter well before the end. These "early" jokes are called jab lines, and any particular narrative joke can have several. Not all jab lines will make listeners laugh, but they're important parts of the joke because they establish a connection with the audience. Consider the following joke that contains several different jab lines, each marked in underline:

*A man wanted to get a pet to keep him company around the house. After some deliberation, he decided on a parrot and chose one that, the sales clerk assured him, was well trained with a full vocabulary of words. He took the parrot home and discovered <u>that it knew quite a lot of words, most of them vulgar, and that it had a bad attitude to match.</u> The parrot <u>spewed out rudeness and vulgarity every time the man entered the room,</u> and the man set out to change the parrot's attitude. He tried repeating nice and polite words around the bird, <u>playing soft music,</u> withholding special treats when it cursed, but nothing seemed to work. <u>The bird just seemed to get angrier and cursed at him even more.</u> Finally, desperately tired of the cursing, <u>he opened up the freezer and shoved the parrot inside.</u> After a few minutes, the cursing and squawking stopped and all was quiet. The man was afraid he had hurt the bird, so he opened the freezer door to check. <u>The parrot looked around, blinked, bowed politely, and recited, "Sir, I am so very sorry I offended you with my language and actions. I ask your forgiveness, and I shall try to control my behavior from now on."</u> Astonished, the man just nodded and carried the parrot back to its cage. As he closed the door, the parrot looked at him and said, "By the way . . . <u>What did the chicken do?"</u>*

In this example, six different jab lines precede the punch line. I challenge you to omit any one of them without softening the joke.

Taking the idea of jab lines a step further, most comics live by what is known as "The Law of Three." This law states that when rhythm is needed to establish the tone or pace of a joke, at least three parts are necessary. As an example, consider this Jon Stewart quip: *I celebrated Thanksgiving the old-fashioned way. I invited everyone in my neighborhood to my house, we had an enormous feast, then I killed them and took their land.* The joke wouldn't be funny without mention of the invitation or the feast because the timing would be off. Humor requires a chance to warm up, and this needed time is very difficult to measure. Perhaps that's why comedians work so hard on the order and theme of their routines—a simple string of jokes without order isn't much of a comedy routine. It's also not likely to build a relationship with the audience.

Like dancing, humor is a form of interpersonal communication, though a complex one. What we find funny depends not just on timing and pace but also on the cumulative build-up of ideas working toward some final point. What sets humor apart from other forms of communication is that it seeks out rules so that it can break them. In our language we expect ideas to be presented clearly. Humor violates that expectation by leading us to believe one thing, then surprising us with the true, intended meaning. We observed this earlier in relation to scripts, and we see it now in relation to communication between people.

Paul Grice is a philosopher of language, best known for developing what he calls the "cooperative principle." First presented in 1968 during a series of lectures at Harvard, the cooperative principle outlines four rules governing polite and effective communication— essentially a guide for proper conversation. For example, Grice's first rule, the Maxim of Quantity, maintains that we should communicate at least as much information as required, but not more. If someone asks me if I know what time it is, and I respond "yes" in an attempt to be funny, this would be a violation of Grice's Maxim of Quantity because, in addition to being a jerk, I'm also providing significantly less information than was requested.

Grice's rules—or maxims, as they are called within the field of linguistics—highlight humor's social nature. They're also useful for identifying the types of social violations exploited by jokes. As an example, consider Grice's second rule, the Maxim of Relation, which holds that our statements must remain relevant, without unwarranted changes in direction or topic. If I tell the joke *How many surrealists does it take to screw in a light bulb? Fish!,* then I'm violating this maxim, because my response has nothing to do with the actual question, despite its implicit statement about surrealism itself. And if I say that *I believe in clubs for young people, but only when kindness fails,* I'm violating Grice's third rule, the Maxim of Manner, which holds that we should always avoid obscurity and ambiguity. Lastly, Grice's Maxim of Quality maintains that we should state only what is definitely true. *What's the difference between in-laws and outlaws? Outlaws are wanted.*

〉〉 〈〈

Some jokes work based on what *isn't* said, thus demonstrating that we can violate more than one of Grice's maxims at a time based on what's left implicit. The joke *How did Helen Keller burn her ear? She answered the phone* initially appears to violate Grice's Maxim of Manner, because the answer doesn't relate to the question at all. But if you know that Helen Keller was blind, then the joke implies additional actions barely hinted at by the answer. In fact, it violates at least three of Grice's maxims by being minimal, irrelevant, and obscure, all at the same time. It doesn't explain why Keller was answering the phone, since she was also deaf, but I suppose that just introduces even more obscurity.

Have you ever e-mailed a joke, only to see it backfire because the humor was lost without tone or context? Of course you have, which brings us to a rare exception—irony. Most humor involves subtle switchings of hidden and intended meanings. By contrast, irony—particularly in the form of sarcasm—isn't subtle. It involves direct confrontations between apparent and secondary meanings. This makes it the only kind of humor where comic delivery is easily measurable.

Studies show that, depending on the nature of the conversation, ironic language either varies widely in pitch or stays completely flat. In other words, when people start speaking ironically, the degree of variation in their tone of voice will either increase or decrease. That may sound difficult to spot, but in fact it isn't—if someone suddenly starts talking differently than a moment before, that person is probably being ironic. Irony shows itself in the face, too. When people use ironic language such as sarcasm, their faces tend to go blank—much like a poker face, except that the expression reveals hidden knowledge rather than concealing it. For this reason, someone speaking ironically can be recognized on a videotape even when the volume is turned off.

Humor is indeed a social phenomenon. It involves the building of personal and social expectations, and when those expectations are broken, funny things happen. Like dancing with a partner and then suddenly giving that person a twirl, a good joke adds spice to ordinary conversation.

## Peer Pressure

*A priest, a rabbi, and a monk walk into a bar. The bartender says, "What is this? A joke?"*

I laughed the first time I heard this joke, not only because it's strange but because it highlights what I call the meta-level aspect of some humor. At first glance it seems to violate all four of Grice's maxims, because there's no communication between the imagined participants. The priest, rabbi, and monk say nothing, so the bartender's comment is surprising. It also seems out of place—unless, that is, you consider that there exists a long history of jokes involving two or more characters entering a bar. This joke relies on that background knowledge, and also on the assumption that the reader expects a joke. In essence, then, the punch line steps out of the context of communication between three clergy members and a bartender, and becomes a comment on the joke-telling process itself.

We call these jokes meta-humor because they involve meta-awareness of joke-telling in general—that is, they take the idea of humor as social violation to a level beyond just breaking polite rules of conversation. In the movie *Wayne's World,* there's a scene near the end where Wayne opens a door revealing a room full of soldiers training for a ninja-style fight. His friend Garth asks what he's doing, and Wayne looks into the camera (movie no-no number one) and says that they're not relevant at all. He just always wanted to open a door revealing a room full of soldiers training as they would in a James Bond movie (no-no number two: the fighters have nothing to do with the plot). That's about as meta as you get.

The idea of stepping outside the normal rules of humor is important, because it shows how comedians don't always speak at face value. When we hear a joke, we build certain expectations, only to have them violated for the purpose of entertainment. This controlled violation allows comics to speak indirectly, conveying messages that might not be appropriate if the topic were addressed head-on. Consider Daniel Tosh, for example. Tosh hosts a clip-based comedy show on Comedy Central called *Tosh.0.* On that show he frequently shares racist and misogynist comments with the television audience, and taken at surface level the comments seem to imply that Tosh himself hates women and minorities. Not so. A closer analysis reveals that Tosh isn't making fun of these groups, he's making fun of stereotypes themselves. For instance, when Tosh urges his viewers to sneak up on women and inappropriately touch their stomachs (ugh!), he's not advocating sexual harassment. Rather, he's satirizing an unusual and inappropriate social phenomenon involving pregnant women, and then exploiting it for the sake of humor.

Still, comics who use this form of humor ineffectively sometimes get themselves into trouble. When Michael Richards (the comedian best known for playing Kramer in the sitcom *Seinfeld*) took the stage at Los Angeles's Laugh Factory on November 17, 2006, he made two mistakes.

First, he lost his cool when a group of unruly fans disrupted his act. They yelled and told him he wasn't funny, and in return he remarked that fifty years ago such behavior would have gotten the African American hecklers hung upside down. Then, as if that weren't enough, he pulled out a few racial epithets and even made an explicit reference to lynching. But his second mistake was even worse—he simply wasn't funny. Had he convinced the audience that such comments weren't his own opinions but a comment on slavery itself, the incident probably wouldn't have been all that remarkable. He might have been accused of going too far, but at least most people wouldn't have considered him a racist, and an unfunny one at that.

Once again we see that humor is a social act. Skilled joke-tellers realize that what's being communicated isn't just a joke but a message about the relationship between teller and receiver. The receiver contributes to the humor just as much as the teller does, bringing a rich set of expectations and—hopefully—a willingness to look deeper than the surface words of the joke.

I must confess that the first time I saw *The Big Lebowski,* I didn't like it. I'd been told by dozens of friends that I *had* to see it; yet when I rented the DVD and settled down on a lazy Sunday afternoon to watch it, I was sorely disappointed. It seemed slow, a little ridiculous, and the humor was sporadic. I know there are worse things to admit, but for me this was difficult because everybody I knew said the movie was great. But then, one evening when a bunch of friends came over for barbecue and beer, I gave in to their urgings and together we gave it another try. I laughed harder than I had in weeks, and *The Big Lebowski* soon became one of my favorites.

As this anecdote shows, humor is sometimes even communal—as with dancing, comedic beats are easier to follow with others around. To see how incongruities and moments of absurdity are shared between people in an experimental setting, let's look at a study by Willibald Ruch, whom we've met several times already. In the field of psychology, experimenters usually remain neutral and in the background. We don't want to influence our subjects because we aren't part of the phenomenon being studied. But what if our goal is to see whether an experimenter's mood can influence the humor ratings of the participant? Could a violation of experimental protocol make a subject laugh more at a joke, simply by setting a humorous tone?

To test humor's social nature, Ruch asked each of his sixty subjects to sit down in a room with a television, at which point a female experimenter explained that they'd be watching ten-minute segments from six successful comedies—for example, Monty Python's *The Meaning of Life,* a particularly absurd series of humorous sketches. Cameras had been set up to record the subjects' reactions to the films, and they were told that afterward they'd be asked to rate their funniness. Immediately following this introduction, the experimenter departed the room so the subjects could view the comedies on their own.

The interesting manipulation occurred after the third film, when the experimenter returned to the room. In the control condition, she sat quietly behind the subject, reading a book and not making any commentary about the movie. However, in the experimental condition she wasn't quiet at all. She remarked upon entering that the next three movies were her favorites, and she laughed audibly at various points during the viewings. To avoid alerting the subject to the manipulation, she made sure not to laugh too long or too obviously.

The results were striking. Subjects who experienced the "humorous experimenter" laughed more intensely and more frequently than those visited by the silent experimenter—almost twice as much. They also judged the last three movie segments to be funnier than the control subjects did, indicating that the experimenter's influence affected not just their behavior but their perceptions too. It was as if the presence of the "humorous experimenter" caused the subjects to like the humor more.

» ««

The influence others have on our subjective moods, particularly humor, is well known. That's why television shows use laugh tracks:

producers know that when we hear laughter, we want to laugh too. As Ruch's study demonstrated, humor ratings can be influenced simply by exposing subjects to laughter. Other studies have found that subjects laugh more and rate jokes as funnier when a nearby actor shares their laughter, and that laughing confederates don't have to be visible so long as they can be heard. In fact, we don't even need to hear the others laughing—simply being told that a friend is nearby and enjoying an entertaining videotape is enough to increase our humorous response.

Shared laughter is more pronounced among people packed closer together . . . we laugh more when surrounded by friends rather than strangers . . . the bigger the audience, the greater the amount of shared laughter. Each of these findings shows that laughter is about more than punch lines. However, they also give the misleading impression that making others laugh is easy.

When we feel as though our emotions are being manipulated by a comic, all bets are off. For example, when we're told that background laughter is canned, rather than live and in response to the humorous material, the background laughter loses its effect. Being instructed to laugh or not to laugh can also backfire. If we're asked to withhold our laughter until after watching a comedy, we're generally able to comply—yet our perceptions of the humor will remain unchanged, as if we had received no instructions at all. This works in the opposite direction, too. If we're told that a joke is hysterically funny and it is, we'll laugh and rate it appropriately. But watch out if an experimenter says a joke is funny and it isn't. Our laughter will disappear, and so will our positive impressions.

In short, laughter isn't contagious like the flu. If it were, we'd never question why others are laughing; we'd only join in. Instead, humor is social the same way our close friendships are social. When shared similarities are explored together, close bonds form. But when laughter is artificial, the result is as satisfying as taking your sister to the prom. It's just not the same.

## Two Brains, One Mind

In my early days as a graduate student I tested a split-brain patient, and I'll never forget the experience. Split-brain patients have literally had their brains cut in half, a procedure called commissurotomy, as a treatment for epilepsy, and the amazing thing is you'd never know their unique medical history from looking at them. The person I tested was named Linda, and the entirety of her corpus callosum had been severed almost forty years before. This is the thick fiber track connecting the left and right cerebral hemispheres, and hers had been cut to stop the spread of seizures that ravaged her brain since young adulthood. Although less radical treatments are now performed to achieve the same effect, the surgery for Linda had been a great success. Despite having two unconnected brain hemispheres, she was not only healthy but nearly indistinguishable from anybody else you might meet on the street. She was sassy and quick-witted and fun to be around—except when asked to participate in laboratory experiments.

The test was an EEG assessment of Linda's brain activity while she performed a "lexical decision" task, which was described in Chapter 5. As you may recall, this task involves showing someone strings of letters and asking if they constitute real words or not. It was a difficult task for Linda because she didn't like experiments, preferring to flirt with the more attractive graduate students, but also because the letter strings weren't shown in the center of her vision. Instead they were presented to the left or right of where she was looking, what we call the left and right visual fields. A curious aspect of the human visual system is that everything we see to the left of our gaze is sent directly to our right hemisphere—and vice versa. The switch occurs because our optic nerves cross sides soon after leaving our eyes, and for most people this medical curiosity has no implications at all, since the two sides of the brain are so highly connected that everything is shared between them right away. But not for Linda. Using this task, I was able to show Linda's right hemisphere both real words and meaningless strings of letters to see if it could tell the difference. It could, even though the right side of our brain doesn't usually contribute to language. That's because the left hemisphere is usually too quick to speak up in normal conversation. To hear from her right hemisphere, I had to look directly at her brain.

Split-brain patients like Linda reveal just how separate the two sides of our brains are, and how divided is our actual thinking. I've seen it for myself—Linda repeatedly assured me that she had no idea what she was seeing, even though her brain recognized the words clear as day. She argued and complained that we were wasting her time with these silly psychological tests, all claims made by her language-dominant left hemisphere. Yet, as I observed in the EEG, her right hemisphere did know what it was doing and was speaking as loudly as it could. It just didn't have access to speech, so that voice had to assert itself in other ways.

Most of us take a whole-brain approach to life, because the corpus callosum allows us to. Nevertheless, the hemispheres of our brain *are* specialized. As we'll soon see, one area of specialization is jokes. Linda had a remarkable sense of humor—and in fact still does at the remarkable age of eighty. But the "humorous side" of her brain has been forced to assert itself in unusual ways, including bawdy humor and obstinacy directed toward struggling graduate students. While Linda was supposed to be looking at words, she would often tell me she was bored and recommended that we continue the experiment on the beach instead. She would ask if I had any liquor stashed away in the lab, suggesting that we skip the experiments to enjoy an early happy hour. Once, she stopped in the middle of an experiment to ask if it hurt when I walked in the sun. It took me several seconds to realize that she was referring to my lack of hair—a condition that hadn't changed since the beginning of the experiment, yet suddenly required immediate attention. Sometimes I think she volunteered to participate in my experiments only because it was so much fun seeing me struggle to keep her on task.

To show how the right side of our brain is so special for humor, I'd like to introduce you to Howard Gardner. Most people know of Gardner as the developmental psychologist from Harvard who developed the popular theory of multiple intelligences, but he's also an active experimenter in several fields, including right-hemisphere humor. It's well known that people with damage to the right hemisphere, usually as a result of stroke, often misunderstand jokes. Gardner conducted a study that tells us why. Specifically, twenty-four subjects—twelve normal control subjects and an equal number of stroke patients, all with damage exclusively to the right frontal side of their brain—were exposed to a series of jokes. However, instead of presenting the jokes in their entirety, Gardner showed only the joke setups, along with four possible endings. Each represented a different kind of conclusion, and only one was funny. Here's an actual joke from the experiment:

> *The neighborhood borrower approached Mr. Smith on Sunday afternoon and inquired: "Say, Smith, are you using your lawn mower this afternoon?"*
> *"Yes, I am," Smith replied warily.*
> *The neighborhood borrower then answered:*

Here came the decision. Four different index cards were shown, each with a different ending. Which one would you choose to complete the joke?

> A. *"Fine, then you won't be needing your golf clubs. I'll just borrow them."*
> B. *"You know the grass is greener on the other side."*
> C. *"Do you think I could use it when you're done?"*
> D. *"Gee, if only I had enough money, I could buy my own."*

Clearly, *A* is the right answer. The other three are valid, just not funny. The second is a non-sequitur ending, meaning that it includes an element of surprise (as a good joke should) but provides no coherence. The third ending doesn't incorporate surprise and is also straightforward. The fourth ending is just sad.

Gardner observed that the patients with right-brain damage had great difficulty finding the correct joke ending, identifying it barely

over half the time. In addition, their errors weren't randomly distributed among the other answers but, instead, favored the non-sequitur
ending. In short, the right-hemisphere-damaged patients could identify that surprise was necessary but had trouble determining what
made the joke actually funny.

From Gardner's study we see an important aspect of right-hemisphere
loss—the inability to identify the meaning of jokes. As we discovered
earlier, every joke involves both spoken and unspoken communication
between teller and receiver. That unspoken communication is what
we need our right hemisphere for. In the example joke above, the unspoken message is that nobody wants to lend something to a neighbor
who never returns things—something the right-hemisphere-damaged
patients missed.

Scientists have been studying the differences between left-hemisphere
and right-hemisphere loss for over a hundred years, but only recently
have we begun to recognize their implications for humor. Damage to
the left hemisphere typically leads to language deficits. If a stroke takes
out the posterior section of our left superior temporal gyrus, we have
difficulty understanding written or spoken language. Loss of the left
inferior frontal gyrus leads to deficits in language production. These
are very different from the effects of damage to our right hemisphere.
Damage to that side of the brain doesn't impair our ability to speak or
understand, but it does affect our ability to connect with people. In
some cases, we experience a muting of emotions. In others, we have a
hard time following conversations or understanding complex aspects
of language such as metaphors. In still others, we lose our ability to
"get" jokes.

Humor doesn't reside solely in the right side of our brain, but it is
certainly right-hemisphere-dominant. This laterality has a big impact
on social interactions because the right hemisphere also helps us recognize the intent behind communication. The main difference between a
lie and an ironic joke is the recognition that the ironic statement isn't
intended to deceive. Right-hemisphere-damaged patients struggle with
ironic humor because they miss this unspoken aspect of the communi-

cation. Normally, we rely on a speaker's gesture and tone to determine if a conversation is sarcastic or ironic. Right-hemisphere-damaged patients don't do this. They function on a literal level, often missing the subtle emotional and nonverbal cues that would otherwise suggest that the conversation is humorous. It takes both hemispheres to fully understand and appreciate a good joke, though apparently they don't need to be connected.

"Humor is preserved in the split-brain patient because both sides remain intact, just separate," says Eran Zaidel, one of the first neuroscientists to study the split brain. His graduate advisor, Roger Sperry, won the Nobel Prize for discovering that each hemisphere can "think" independently. "I've seen them [split-brain patients] display some marvelous senses of humor too, telling jokes all day," Zaidel continues. "But because skills like maintaining social relations are specialized for the right hemisphere, while language is lateralized to the left, that makes humor sometimes harder to spot. It becomes very important how you look for it, and how you allow it to come out. You can't look only at the words."

On one occasion, Zaidel counted fourteen different kinds of jokes told by Linda and a second split-brain patient, Philip. The difference between their jokes and those of the general public, however, is that theirs are less tied to language, which resides in the left hemisphere. This is especially true of the jokes told by Linda, who seldom uses puns or other wordplay. However, she excels in social humor, especially the kind that teases—including herself. "I told my husband I'm a lot smarter than him," she once said. "I have two brains and he only has one."

We often don't appreciate how much the hemispheres work together to provide us a full cognitive experience. This has implications not just for humor but for consciousness itself. Zaidel once asked Philip a series of questions directed to his left and right hemispheres and found that the two sides of his brain had different personalities and outlooks on life. His left hemisphere experienced relatively low self-esteem while his right hemisphere saw itself rather positively. The right side also experienced greater loneliness and sadness. Another split-brain patient's right hemisphere was particularly influenced by childhood memories of being bullied, even though his left hemisphere denied finding such experiences disturbing. And then there was the split-brain patient who, when asked if he believed in God, responded "yes" with his left hemisphere and "no" with his right.

This division of resources in the brain has strong implications for how we think. For example, though the left does most of the heavy lifting when it comes to language, the right contributes understanding in the form of recognizing subtleties, including those in jokes. This suggests that the right hemisphere is important for coming up with insightful connections. It's also important for poetry. When poetic language in the form of creative metaphors is shown to the right hemisphere, we're a lot better at processing that language than when it's shown to the left. So, perhaps the right hemisphere is like our nonliteral friend, flitting from topic to topic, helping us with poetry, jokes, and other artistic endeavors. Alone it would be lost, but when paired with the stricter and more literal left hemisphere it provides us just the balance we need to remain insightful and creative. Without either hemisphere we might be lost, but with the two combined we have a powerful ability to understand and create.

## FUNNY RELATIONSHIPS

As a social phenomenon, humor has a direct impact on our relationships. As we've seen, being around laughing people increases the chance we'll find a joke funny. But the influence works in reverse, too: enjoying a humorous attitude improves the quality of our social relationships. This reveals something important not only about humor—that it brings us closer together by providing shared experiences—but also about relationships themselves. We bond with people who share similar perspectives toward life. Humor is the best way to uncover what those perspectives are.

We don't need to look hard to find scientific proof that humor is important to romance. Numerous researchers have asked people what traits they most desire in a partner, and one trait is always near the top of the list: sense of humor. A 2007 study published in the journal *Archives of Sexual Behavior* found that sense of humor was the second most desired trait, behind only intelligence. Women ranked it first. For men it was number three, after intelligence and good looks.

However, this affinity for humor hasn't always been so strong. In a similar survey taken in 1958, humor ranked much lower among women's preferred traits for mates, after such characteristics as "well groomed," "ambitious," and "makes sensible decisions about money." In 1984 it ranked behind intelligence and sensitivity. And in 1990 it was number two—again, behind sensitivity.

One possible reason for this shift in priorities is that women, because they're no longer confined to a limited number of jobs, have begun to expect different things from their men. Ambition and money management abilities are important in a partner, but they become a lot less relevant when these responsibilities are shared in the partnership. A strong and ambitious man is nice, but better still to find one who is funny too! But this still doesn't answer: What's so special about humor?

Before exploring that question, we need to recognize that affinity for humor isn't universal; it's part of our culture. For example, humor almost never fares as well in other countries. In a survey of Siberian women, humor didn't appear even among the top-ten most important traits in a partner. In fact, it was closer to twentieth. Perhaps this says something about women in Siberia, but I think it says more about women in the United States. In the United States, we want to have fun, to enjoy ourselves, and be entertained. This desire isn't superficial but an important part of relationship building. In Siberia people also want to have fun, but faithfulness (#2), reliability (#4), and love for children (#9) are all more important because life in Siberia is hard. Russians are a jovial, generous people, but let there be no mistake—when food is scarce and snow and vodka are plentiful, having a spouse you can depend on to help maintain the home is invaluable.

Perhaps humor is so important, especially to American women, because it evolved to be over time. It helps us convey our thoughts and values, two important goals for identifying compatibility, and it also helps build social bonds. From an evolutionary standpoint, these benefits raise some interesting questions. For example, could humor have evolved to predict quality of mates? Is there something special about humor that singles out funny males as especially good partners?

Understanding how natural selection brings about any complex behavior, including humor, is difficult because it involves speculative story telling. It's like seeing a pool table filled with moving balls and guessing the direction and velocity of the strike that got things going. Yet, it's still useful to guess why humor became so important for our species. We'll never know for sure why it evolved the way it did, but scientists have some good theories, and they say a lot about humor differences between the genders.

The evolutionary argument starts with the premise that women have more at stake in procreation because they have so few opportunities to birth children. Each attempt, if successful, requires at least a dozen years of nurturing. Their opportunities also end late in middle age, meaning that a woman might not get many tries, so each one has to count. By contrast, men can father multiple children simultaneously, almost up until death, and they don't have to stick around after their initial contribution. So while men can be less discriminating, women must be selective and use subtle lures to attract only the best of mates. Laughter is one of those lures, just as sense of humor is one of the ways males show their suitability. The argument makes at least two predictions: that women should laugh more (indeed, as we've seen, they laugh roughly 125 percent more than men) and that humor should play different roles for men than for women. For men, the ability to be funny and make a partner laugh should be the most important consideration. For women, it should be the ability to appreciate humor.

Indeed, these predictions appear to be true. One study conducted by the psychologist Eric Bressler at Westfield State College in Massachusetts asked male and female subjects what was more important: having a partner who's funny and produces quality humor, or having a partner who appreciates one's own jokes. This question was asked as it applied to several kinds of relationships, ranging from one-night stands to long-term romances. The results were clear—in almost every relationship category, women preferred men who were funny, and men preferred women who appreciated their own humor. The sole exception was platonic friendships, the only kind for which offspring are impossible (assuming they remain platonic). For that category, men didn't care whether they were the funny ones or not.

Regardless of what we believe about its evolutionary history, humor keeps us healthy—both mentally and physically. It makes us more desirable by revealing either our openness to laugh or our dedication to bring out laughter in others. This may explain why people who rate high on tests of intimacy also have a good sense of humor. The same goes for trust, dependability, and kindness.

In short, humor is key not only in mate selection but for maintaining healthy partnerships too. Relationships take work, and an excellent way to spot a mind willing to put in the effort is to look for a good sense of humor. Nine out of ten couples say that humor is an important part of their relationship. Compared to those in dysfunctional marriages, couples in strong ones also say that they value and appreciate their partner's humor more. Indeed, studies examining long-term couples—those who have remained together for forty-five years or more—have found that laughing together is essential for marital success.

Humor appears to be as important for establishing healthy relationships as it is for maintaining healthy bodies and minds. Just as a humorous attitude signals an engaged mind, a shared appreciation for humorous living signals a fit partnership or marriage. A good sense of humor is more than a perspective or outlook. It's a means of sharing expectations with someone close to us.

So, humor dances—and there's no better way to build a solid relationship than finding somebody who dances to the same rhythm.

# 8

## » OH, THE PLACES YOU'LL GO

*Men will confess to treason, murder, arson, false teeth, or a wig. How any of them will own up to a lack of humour?*

—FRANK MOORE COLBY

THIS FINAL CHAPTER OPENS WITH A STORY, AND AN UNUSUAL one at that. It's the tale of an arm-wrestling match between two CEOs, held in front of a huge audience to determine the ownership of an advertising slogan. It's an unusual turn for a book on humor, but it shows how humor is everywhere.

It isn't every day that major companies settle legal disputes over an arm-wrestling match, but Stevens Aviation and Southwest Airlines aren't your typical companies. The event in question started when Southwest began using the phrase "Just Plane Smart" in its advertising campaign. The slogan matched Southwest's smart and irreverent personality and was a big success, except for one problem. Stevens Aviation, an aircraft maintenance firm based in South Carolina, was already using it. Actually, Stevens Aviation was using "Plane Smart," but the two slogans were close enough that lawyers quickly became involved. Disputes like this are common and usually end with one side giving up its slogan, but Southwest's CEO Herb Kelleher had another idea—he personally challenged Stevens's CEO Kurt Herwald to an arm-wrestling match. Company employees would be the spectators, and all money raised by the event would go to charity. The winner of the match would keep the slogan, while the loser would explain to his board of directors why he lost the rights. It was the kind of challenge no smart businessman turns down—especially one like Herwald, who was young, athletic, and an avid body builder.

>> <<

To appreciate the audacity of the challenge, you should know that Kelleher is the opposite of a body builder. In training videos he flaunted his pudgy, nearly forty-year-old body, softened by alcohol and tobacco. Cigarette in mouth, he trained for the match by lifting bottles of Wild Turkey whiskey. It took three stewardesses to help Kelleher complete a sit-up in his videotaped preparation for the match.

The "Malice in Dallas," as it came to be known, was held on a sunny morning in late March 1992 at the Dallas Sportatorium, in front of hundreds of fans. Crowds chanted "Herb! Herb! Herb!" as Kelleher arrived, his gut barely covered by a poorly tied bathrobe. Kelleher's right arm hung in a sling due to an injury he got "while saving a little girl from being hit while running across the I-35 freeway." He also complained of suffering from a weeklong cold, as well as athlete's foot, but this didn't stop him from lunging at Herwald as he entered the ring. Officials had to restrain both CEOs.

"We don't need no stinkin' lawyers, we're going to do this like real men. In the ring," called the announcer. The fight was on.

In a business world where publicity can be a company's biggest asset, the event was already gold. Hundreds of people had come to watch the event, and so had dozens of television stations, including CNN and the BBC. Making sure to put on a show, Kelleher started by presenting a substitution order from the supreme court of Texas. In his place would be Texas professional arm-wrestling champion J. R. Jones. Herwald objected, but officials ignored his complaints—and Southwest won the first of three matches. At this point, Herwald announced, "If they can bring in a ringer, I can bring in a ringer," and brought out his own substitute. But instead of calling upon a professional, Herwald introduced "Killer" Annette Coats, a tiny Stevens Aviation customer representative who weighed maybe half as much as Kelleher. Still, she won the second match handily.

By this time the event had turned into mayhem. Herwald ended up beating Kelleher in the third and final match, but not surprisingly Kelleher protested and things started to get strange. For reasons not entirely clear, a professional wrestler jumped into the ring and began choking Kelleher, and as Kelleher fell to the ground Herwald returned to the ring to defend him from the muscle-bound man in tights. A brief scuffle ensued, and in the end Kelleher and Herwald chased away the spandexed intruder, finally ending the dispute with a shake of the hands.

"Just to show there's no hard feelings, or to be accused of taking advantage of senior citizens," Herwald announced as things settled down. "We've decided to allow Southwest Airlines to continue the use of our slogan. *Our* slogan. In exchange for a $5,000 contribution to the Ronald McDonald house, which needs the money more than Southwest Airlines does."

》. 《

The event was indeed a publicity coup. It made Southwest and Stevens appear hip and funny and established Kelleher and Herwald as confident managers, willing to play the fool for the sake of their companies. When Kelleher was interviewed after the event—while sitting on an ambulance stretcher, of course—he was asked how much Southwest would normally have paid for such advertising exposure. "Why, I never even thought about it in those terms," he replied tongue-in-cheek. The president of the United States wrote Kelleher two days later to congratulate him on the brilliant idea. *BusinessWeek* and the *Chicago Tribune* wrote that Kelleher and Herwald's willingness to set aside stuffy business images to provide entertainment was one of the things that made Southwest so special. In fact, it was the only airline to make a profit in every one of its thirty-one years.

In this chapter, we're going to take a different approach to humor. It's good to know that our brains are conflict-processing machines, turning things like ambiguity and confusion into pleasure, but for those of us who just want to tell a good joke, it's time to look at humor as it is applied in the real world, seeing how people like Kelleher and Herwald play funny to their advantage. Humor makes us better workers, students, and managers, and it's important to recognize how it's used in each of these settings. It's also important to learn what science has done to help us maximize our humor potential.

## OH, THE PLACES YOU'LL GO

Something about the business world changed in the 1980s. Southwest Airlines was no longer the only company using humor in the workplace; companies all over the world began to recognize that humor sells. For instance, the newly hired president of New England Securities introduced himself to new employees by reading from the Dr. Seuss story "Oh, the Places You'll Go!" as a way of promoting the company's core values. The computer manufacturer Digital Equipment created a "grouch patrol" to roam the workplace and identify sour workers. And the San Francisco Police Department hired humor consultants to update its Neighborhood Crime Prevention workshops after learning that previous workshops had left residents *more* traumatized by crime, not less.

At a minimum, we know that such approaches improve employee morale. A survey by the Campbell Research Corporation found that 81 percent of the companies instituting programs like "casual Fridays" experienced improved mood in the workplace, and half saw concomitant improvements in productivity. Indeed, the idea that a laughing worker is a lazy worker became as outdated as the notion that being serious means never laughing. "No subject is ever too serious for humor," says John Cleese, Monty Python member and founder of Video Arts, the largest producer of corporate training films in the world. "I think many people have a basic misunderstanding: There's a difference between being serious and being solemn."

As it turns out, the difference between seriousness and solemnity is quite important in the professional world. Seriousness keeps us focused on improvement, and that's undoubtedly beneficial. Solemnity often achieves the same thing, but it does so by emphasizing formality and avoiding cheer. That's fine too, especially for certain situations where cheer would be inappropriate. But there are times when a bit of joviality is needed too.

Fortunately science *has* shown that joviality helps us at our jobs, and a wide variety of jobs at that. For example, being funny helps hide flaws in our organization skills. Most of the time when we give speeches, we're careful to organize our points in a logical and meaningful order. But studies have found that we can give those same speeches with points mixed at random, and as long as we also incorporate humor, viewers won't notice. When a healthy dose of jokes and humorous anecdotes are included among a jumble of ideas, the results can be just as informative as if they were well organized.

Humor is important in the educational world, too. One of the most studied humorous environments is the classroom, where finding after finding shows that students prefer taking classes with fun teachers. Humor makes classroom environments more enjoyable, increases student motivation to learn, and leads to more positive evaluations of teachers. Take, for example, educational consultant Bill Haggard. When Haggard's students began having trouble turning their homework assignments in on time, rather than increase punishments he developed a three-way excuse chart. The three sides corresponded to the three most common categories of excuses—*Helpless, Hopeless,* and *Not in Control of Body*—and every time a student had a problem, the class worked together to decide its proper location on the chart. As the year went on, excuses dwindled and students started taking responsibility for missing their deadlines. "Teachers are way ahead of the game if they can change anxious situations into humorous ones that evolve into shared experiences," Haggard said.

This approach is useful for students of all ages. It has even worked at conservative schools like the United States Military Academy at West Point, where students were asked to judge particularly effective leaders in terms of their humor, as well as other characteristics such as physical ability, intelligence, and consideration. Good leaders were rated significantly more humorous than bad leaders, even when these other variables were controlled for.

These findings suggest that humor makes the classroom more fun, but does it actually help learning? Absolutely so. Consider, for example, a study involving more than five hundred students at San Diego State University. The students were enrolled in what they thought was a normal Introductory Psychology course on Freudian personality theory, but different students attended different kinds of lectures. One lecture incorporated humor relating to the course content. A second lecture included humor that wasn't related to the material but still kept students entertained. And a third lecture used no humor at all, only a serious treatment of the subject material. When the researchers tested students' retention six weeks after the lectures, they found that those who attended the lectures with humor related to course content scored significantly higher than the other students. In short, humor is beneficial to learning—but only when it focuses on what we're trying to learn.

Connecting humor with learned material is key because it keeps the mind focused. Humor forces our minds to work more than if ideas are presented in a straightforward manner. This work is essential for the same reason that lifting heavy weights builds muscles—because the extra effort makes us stronger. The benefits of humor even last for long periods of time. Take for example the following cartoon, which was used to teach children statistics at Tel Aviv University. It depicts an African explorer remarking to a group of children not to worry about local predators: "There is no need to be afraid of crocodiles," the explorer says. "Around here their average length is only about 50 centimeters." In the background is a huge crocodile coming to eat the explorer, and the children mutter that he, too, shouldn't forget about standard deviations.

When teachers were trained to use such jokes in their classrooms—even as few as three per lesson—learning increased by almost 15 percent. These improvements lasted throughout the entire semester.

These findings aren't lost on professionals who frequently speak in public. Congressional debates, Supreme Court hearings, White House briefings—each of these venues is rife with humor, especially when it helps tackle difficult topics. Joe Lockhart, press secretary for President Clinton from 1998 to 2000, was a master of humor. On one occasion he was asked to explain the frequent foreign-relations trips being taken by the First Lady. At the time, such trips were a significant expense and getting more attention than the administration wanted, and Lockhart replied that they were paid for by the State Department. "Joe, is there a difference between 'the State Department' and 'the taxpayers'?" asked a member of the press corps. This was a serious question, since government spending was a hot political issue. Lockhart's only solution was to turn the situation into a joke, and it worked. "No," he replied. "It just sounded better if I said 'State Department.'"

From these examples it may seem that humor is merely self-serving, hiding organizational flaws and diverting attention away from unwanted topics. In a sense it is, but it can serve broader purposes too. In politics, especially, humor can be an invaluable weapon. Everybody over thirty remembers the debate between Lloyd Bentsen and Dan Quayle during the 1988 presidential elections. Both were running for vice president, and in response to a question about his ability to fill in for the president if needed, Quayle had just compared himself to the late John F. Kennedy. Bentsen would have none of it.

> QUAYLE: *I have as much experience in the Congress as Jack Kennedy when he sought the presidency.*
>
> BENTSEN: *Senator, I served with Jack Kennedy. I knew Jack Kennedy, Jack Kennedy was a friend of mine. Senator, you're no Jack Kennedy.*

The audience members laughed and cheered for so long that the moderator had to interrupt and calm them down. With just a few words Bentsen had crushed his opponent, but he avoided appearing cruel because there was a bit of playfulness in the comment too. Talk about multiple messages—Bentsen simultaneously compared two politicians, affiliated himself with one and not the other, and implied that one of them was a child in need of rebuking.

Yet like any other weapon, humor can backfire too. One study asked more than a hundred potential voters, immediately before the 2004 presidential elections, to imagine a scenario in which two fictitious politicians get into a heated debate. The debate becomes so intense that a moderator is forced to intervene, at which point one of the politicians apologizes for his exuberance. Here are two versions of his apology, and I want you to guess which one subjects considered more effective.

> *"I know I can go a bit overboard when I get going. My daughter even said that maybe the moderator should have music to drown me out when I go on too long, like they do at the Emmy Awards."*

> *"You must excuse me, but when I am right, I get angry. My opponent, on the other hand, gets angry when he is wrong. As a result, we are angry at each other much of the time."*

The first one was rated more effective, by far. It was judged better at improving the debate, overcoming conflict, and finding common ground. The preference for humility was so strong that it even overcame political ties. When the candidates described in the scenario were given labels such as "Democrat" and "Republican," the first remark was deemed funnier and more effective even when the person judging belonged to the opposite political party.

In short, political humor doesn't have to be insulting to succeed. It doesn't have to support one's existing values and beliefs, either. It simply needs to reveal what one is thinking, preferably with humility, so that others can join in.

Bringing people together is indeed a prominent theme in this book—humor clearly builds bonds within social groups. One professor of management found that joking personalities among New Zealand information technology employees improve workplace dynamics by safely allowing the employees to question authority. A year-long ethnography study of hotel kitchen workers showed that humor, even when critical, solidifies groups by highlighting shared beliefs and responsibilities. And an in-depth observation of Sardinian fish market workers revealed that humor brought workers closer together by reminding them of their common goals.

Many groups have their own brand of humor, each with its own style. Jewish humor is one of the oldest. *"The Jewish people have observed more than 5,758 years as a people, the Chinese only 4,695. What does this mean to you?" asks the Rabbi. "It means the Jews had to go 1,063 years without Chinese food," replies the student.* Lesbian humor is popular too, with one journal article—"How Many Lesbians Does It Take to Screw in a Light Bulb?"—eliciting so much controversy that more than forty pages were needed to resolve it. There's even a category of jokes called "white trash humor." These jokes often start with the phrase "You might be a redneck if . . . " (as popularized by Jeff Foxworthy in the 1990s), and according to the linguist Catherine Evans Davies, they legitimize Southern working-class individuals by separating them from lower social classes. "Redneck" refers to respectable working-class people. "White trash" does not.

In short, humor exercises the mind, which in turn makes us better students and teachers. It also allows us to appear more organized and helps us get our points across, whether in the courtroom or the fish market. In the next section we'll address an even more important aspect of humor—its relationship to intellect. Intelligence and humor both involve "thinking messy," and now it's time to see how training the brain to be funny is just another way of training it to think smarter.

## GREATER IMPLICATIONS

Humor is complicated because we, ourselves, are complicated. We laugh, and cry, and have malleable personalities because our brains have developed over generations to be adaptable. Without the ability to laugh, we wouldn't have a way to react to much that happens to us. Without having a sense of humor to take pleasure in the incongruous or absurd, we might spend our whole lives in a perpetual state of confusion, rather than occasionally transforming those feelings into amusement.

In this sense, humor is as important an evolutionary trait as intelligence, because without it we wouldn't be able to cope with the complex world we've created. As discussed earlier, humor evolved over successive generations, much like our ability to use tools and language. Humans need a way to deal with conflict and confusion, and what better way to do that than laugh? Like creativity and insight, humor has allowed us to solve our problems without resorting to beating each other over the heads with sticks. And as a fundamental part of who we are, humor has developed a close relationship with each of these other uniquely human abilities.

Amazingly, humor is correlated with IQ even by the age of ten. This finding was observed by the psychologist Ann Masten, who showed a group of ten-year-old children a variety of Ziggy cartoons chosen for their varying complexity and sense of humor targeting this age group. As the children rated the cartoons subjectively and explained why each one was funny, Masten recorded their faces, noting instances of smiles or laughter. She then showed the children a series of captionless cartoons and asked them to come up with humorous titles. The children's ability to correctly explain the cartoons was used to determine their "humor comprehension" while their ability to come up with funny captions measured their "humor production."

Masten found that both humor comprehension and production were significantly correlated with the children's intelligence, which she had also measured separately. For comprehension the correlation was 0.55, and for production it was 0.50—very large numbers, considering that the maximum possible correlation is 1.0. Even the extent to which the children laughed at the cartoons was closely linked with their intelligence, with a correlation of 0.38. Given that IQ has about the same correlation with job performance, humor probably predicts intelligence as well as do most practical measures of life success.

As noted earlier, learning to be funny may even make us smarter. Consider the following scenario. You enter a laboratory and are told to complete a problem-solving task. But before you do, you must watch a compilation of funny bloopers from popular television shows. Other subjects aren't so fortunate—they have to watch a five-minute documentary on Nazi concentration camps. Still others see a math film titled "Area Under the Curve." And a fourth group doesn't watch a film at all but, instead, is asked to eat a candy bar and relax, or to complete two minutes of exercise by stepping on and off a concrete block.

Each of these manipulations is intended to affect mood in a different way. The math film is expected to have minimal impact, and the Nazi concentration camp documentary should be depressing. Both the candy and the funny video are expected to induce a good mood, but only one is meant to elicit laughter. The question is: Can laughter alone influence performance on the task that follows?

To find out, a final task is administered, and it's challenging too. It's called the Drucker candle insight task, and it goes like this: You're given a box of tacks, a candle, and a book of matches. Then you're asked to attach the candle to the wall so that it burns without dripping wax on the floor. The solution (as you might already know from previous exposure to this puzzle) is to attach the empty box to the wall using one of the tacks and then to use wax or another tack to secure the candle atop the box. What makes this task challenging for many people is "functional fixedness"—the inability to view the box as serving any purpose other than holding tacks. The candle doesn't have to be directly "attached" to the wall. And boxes can do more than just hold small objects.

When this experiment was conducted by Alice Isen and two colleagues at the University of Maryland, only 32 of the 116 subjects came up with a solution. But when the results were analyzed based on what the subjects did before trying, an amazing finding emerged. Only 2 of the subjects in the math-film group solved the puzzle. Only 5 did from the exercise group. In fact, no group of subjects performed better than 30 percent—with one exception. The subjects who were shown the funny bloopers succeeded at a rate of 58 percent (11 out of 19 subjects).

When I told my wife about this finding, she asked why I wasn't a genius by now. After all, I've seen hundreds of comedies in my life, so why hasn't that made me the smartest man in the world? It was a good question. My reply was that I'm not a genius, but imagine how stupid I'd be if I hadn't watched so many *Fawlty Towers* reruns. It was the best comeback I could come up with.

Insight isn't the only complex cognitive skill that benefits from humor. Another is "mental rotation," the ability to rotate objects in our heads—a common task for assessing spatial ability. As it turns out, people who are presented with funny jokes are faster at turning and twisting abstract shapes in their minds, even when the jokes involve minimal visual imagery. Reading funny jokes also improves our scores on creativity tests, reflecting increased mental fluency, flexibility, and originality. One study even showed that watching videos of Robin Williams's stand-up helps us come up with unlikely solutions for word-association problems.

It's hard to say why watching comedy makes us smarter and more creative. Perhaps, by being exercise for the mind, humor provides a much-needed warm-up. As I shared before, in my life I've run two mar-athons, and both times I was in fair shape. But that was a while ago, and a lot has changed since then. Now if I tried to run that far I'd prob-ably collapse into a fetal ball. Exercise doesn't change us forever, and neither does humor. As a form of mental exercise, humor keeps our brains active. Our brains must be exercised regularly, and when they are—well, we become capable of just about anything.

## BECOMING FUNNY

In 1937 fewer than 1 percent of respondents admitted to having a below-average sense of humor. Forty-seven years later, a similar sur-vey showed that 6 percent of people admitted to being sub-par in the funny department. So, the question is—are people getting funnier, or are our self-concepts becoming slightly less deluded?

A rudimentary understanding of statistics reveals that the first explanation can't be true, because by definition at least half of us must be below average. I like to call this the "Dane Cook Effect": no matter how funny we think we are, we're probably being optimistic. It's easy to believe we're funny when our spouse or mother laughs at our jokes. But the truth is, being funny is hard. If it were easy, everybody would be a comedian, even Dane Cook. I don't mean to imply that Cook is an unfunny individual. I've seen him perform, and I enjoy his stand-up immensely. Cook has earned millions of dol-lars from movies and comedy specials, yet his success hasn't stopped him from being openly disliked within the comedic world—mostly because he's a master storyteller, not a comedian. His performances are certainly entertaining, but they're based on anecdotes, not on humor—like Lenny Bruce except without the edge. Perhaps this is why a tournament of "The Sixteen Worst Comedians" held in Boston, Cook's home town, rated him "the worst of all time." *Rolling Stone* once made a list of things funnier than Cook that included prune Danishes. Comedy is hard.

Sadly, textbooks can't teach us how to be funny like they teach us calculus. It's simply too complicated for that (even compared to calculus). But there are a few things that aspiring comedians ought to know.

Fortunately for people with unfunny siblings, there appears to be little genetic influence over humor. So it doesn't matter who our relatives are, because everybody has an equal chance of being funny. Scientists know this by comparing identical and fraternal twins, as well as biological and adopted siblings. Intelligence has a heritability of 50 percent, meaning that half our smarts are determined by our parents. Height has a heritability of over 80 percent. By contrast, for humor that figure is probably less than 25 percent.

Much of what we know about the comic personality comes from a book by Seymour and Rhoda Fisher, psychologists at the State University of New York at Syracuse. The book is titled *Pretend the World Is Funny and Forever,* and over the course of 288 pages it explores the pathways taken by more than forty professional comedians, luminaries such as Woody Allen, Lucille Ball, and Bob Hope. Through interviews, observational study, and background research, the authors break down the personality characteristics of these successful comics, looking for patterns in the life experiences that made them funny.

The authors found that fewer than 15 percent of the comics thought they would be professional humorists when they started out—evidently, it's never too late to think about getting into comedy. Most received little support from their parents. Many had been class clowns. And each expressed his or her humorous perspective in a unique way. Some were lavish and highly expressive, like Jackie Gleason. Others were quiet and reserved, like Buster Keaton. Still others were social and spontaneous, like Milton Berle, or reclusive, like Groucho Marx. It seems there are almost as many ways to act up and be funny as there are comedians. But they all had one thing in common: a deep interest in sharing observations with others.

"The average comedian moves among his fellows like an anthropologist visiting a new culture," write Fisher and Fisher. "He is a relativist. Nothing seems natural or 'given.' He is constantly taking mental notes."

The idea that comedy involves observation isn't new, though it's still important. Humorists question everything they see, never taking anything for granted. They tell jokes and humorous anecdotes because they feel compelled to share what they see. Fisher and Fisher saw this desire in their interviews and even during psychological assessments like the Rorschach inkblot test. That involves looking at inky, amorphous blobs and describing what they look like, and when comics looked at these blots, rather than provide simple interpretations they consistently turned them into stories. Frightening wolfmen weren't evil, just misunderstood. A pig-like face wasn't ugly, it was endearing. A blot that resembled the devil was interpreted by one comic as silly, even goofy.

These observations show how an active mind is a humorous mind, and that the more we keep our brains working, the more our humor benefits. Consider also this important fact—maintaining a humorous attitude, as measured by the ability to recognize humor when it presents itself, is strongly related to actually *being funny*. I'm referring to a study conducted by the psychologists Aaron Kozbelt and Kana Nishioka in which subjects were asked to identify the meaning and content of funny cartoons—a measure of humor comprehension. Note that this is very different from appreciation. Appreciation was measured too, but here I'm talking about how well subjects understood the cartoons, a matter of recognizing the source of the jokes' incongruity. The researchers also measured humor production by asking subjects to come up with funny captions for an entirely different set of cartoons. Independent judges then rated how funny those captions were.

No significant relationship was found between humor appreciation and humor production, meaning that merely liking humor—as in,

*"Never, ever, think outside the box."*

FIGURE 8.1. A cartoon used to explore the link between humor comprehension and production. The ability to recognize that the man is warning against inappropriate feline creativity suggests an ability to actually *be funny.* Cartoon by Leo Cullum, www.cartoonbank.com.

enjoying a good laugh—doesn't make us funnier. Rather, what matters is how well we understand the mechanisms behind the jokes. Consider Figure 8.1 as an example.

If your interpretation is that the man is warning the cat against being "inappropriately creative," then you understood the cartoon correctly. You are also more likely to produce funnier jokes, as found by the authors of the study. Specifically, those who score high on recognition also score high on production, even though subjective appreciation has no effect. In short, simply understanding jokes makes us funnier people.

If you thought that the humor stemmed from the man foolishly talking to a cat that doesn't understand English, you might want to buy a few joke books, or maybe get a subscription to *The New Yorker* and study a few more of its cartoons.

There's no shortage of companies willing to improve customers' sense of humor. For example, a workshop run by the comedian Stanley Lyndon promises that readers of his book will produce jokes 200 percent funnier than before. And an online course conducted by the ExpertRating training company offers to certify individuals in online humor writing, for a mere $130, as a way of preparing them for the lucrative career of comedy-club performance. From these programs, it might seem that learning to be funny is easy. It isn't. In fact, as discussed in this book's introduction, there's only one proven way to improve one's humor: namely, by following the *Rule of the Five P's:* practice—and practice—and practice—and practice—and practice.

I'm going to conclude this chapter with one last study, this time by the Israeli psychologist Ofra Nevo. She wanted to know what makes people funny, but rather than giving personality tests or administering surveys, she put groups of teachers through a seven-week course on improving humor—twenty hours of training in all. Specifically, her aim was to find out if simply learning more about humor was enough to make people funnier to be around.

First, Nevo grouped her one hundred training subjects into several different experimental conditions. Some received an extensive humor training program, with numerous exercises providing background into the cognitive and emotional aspects of humor. They practiced telling jokes in front of larger groups. They talked about different humor theories and styles. They even explored its physiological and intellectual benefits—much as we've done in this book. Others received similar training but without the practice—a passive version of the humor training program. Still others received no training at all. All of the subjects took a humor assessment test at the beginning and end of the experiment, and received a questionnaire asking how helpful they thought the training was.

Nevo found that, on average, the subjects didn't find the training very helpful. They rated the program effectiveness between "small" and "medium"—the equivalent of between 2 and 3 on a 5-point scale. She was disappointed by this result, but, as we'll soon see, perhaps she shouldn't have been. After the training was over, peers were asked to rate the subjects on their humor production and appreciation, with questions like "How much is this person able to appreciate and enjoy humor produced by others?" and "How much is this person able to create humor and make others laugh?"

As it turns out, the subjects who took the training scored significantly better on both measures, improving by up to 15 percent. Even those who hadn't practiced the techniques showed some progress. In short, although the subjects themselves thought they hadn't become any funnier, the people around them disagreed.

It's tempting to ask the obvious question: Only 15 percent? That hardly seems like much, especially compared to Stanley Lyndon's promise of 200 percent. But imagine how great it would be to be 15 percent smarter. Or 15 percent more attractive. If I were 15 percent taller, I'd be close to the size of an average center in the NBA. I'll take 15 percent any day.

Sadly, not much follow-up research has been conducted on Nevo's work, so we still don't know why the subjects in her experiment didn't "feel" funnier, even though their peers believed they were. One possible explanation is that sense of humor is a trait, as discussed earlier. Traits don't change quickly, meaning we'd need more than twenty hours of training to see obvious improvement. It might also be that changes in humorousness are subtle, more so than can be detected with scientific measurements. This might help explain why professional comics with years of experience often claim that they've only begun to learn their craft. Humor isn't something that can ever be mastered. It can only be learned.

And that learning occurs over a lifetime of practice—and practice—and practice—and practice—and practice.

# CONCLUSION

*It is more enjoyable to read a humorous book
than to read one explaining humor.*

—AVNER ZIV

"I'VE GOT A SPOT. I'M GOING ONSTAGE NEXT SUNDAY."

That's what I told my wife Laura after finishing this book. I had
signed up to perform a short act for amateur night at a local comedy
club, and although I was terrified, it seemed the right thing to do. I
had spent over a year of my life reading article after article, book after
book on the topic of humor. My mind would never be more tuned to
understanding, analyzing, and dissecting what makes jokes funny. If
there was a time to apply that knowledge, this was it.

"I thought the book was supposed to be about science," Laura
replied. "Not a how-to book."

Laura was less enthusiastic about my proposal than I had hoped. Still,
I didn't blame her because she was right. The book wasn't intended to

be a practical guide, and I had no interest in comedy as either a hobby or profession. But I still felt like I needed to apply what I had learned in a real-life setting. Just as an art professor would never teach a class without having slapped a brush against canvas, I had to see what "doing comedy" was like. Didn't I?

"I need to at least try what I've been talking about. I might not be any good, but I want to know what it feels like to tell a joke in front of an audience."

Laura stared at me with blank eyes, as if I'd just said I was taking up professional baseball because I knew the batting stats of the entire Red Sox lineup. There's a difference between understanding what makes a joke funny and being able to share that joke in an entertaining way. Laura recommended I take an improv class first, which I agreed was probably a good idea. But that would also defeat the point. I wasn't actually trying to be funny, at least any funnier than I already was. I simply wanted to know what performing felt like. I wanted to know where to draw the line between science and art, so that I could share just how far knowledge of incongruity and surprise can take an aspiring humorist. I wanted to know where theory ended, and where fluid performance began.

In short, I wanted to get in over my head. And I did.

The night of the performance I was nervous, of course, but I comforted myself by practicing the act I had written down on 3×5 cards. Because the show was held on a Sunday night and only amateur performers were taking the stage, I expected the turnout to be small. I was wrong. With only a small cover charge and cheap food and drinks, the comedy club was packed. I wound up being the third person to perform that evening, and by the time I stepped onto the stage I was seriously regretting my decision. But it was too late to change my mind.

"How is everybody doing tonight?" I asked as I took the microphone and looked out onto the crowd, knowing I had to say something to get the show started. As expected, many audience members cheered, fueled by cheap beer and buffalo wings.

I froze. After all the late nights I'd spent in libraries trying to find sixty-year-old articles about people's reactions to Benny Goodman albums, I was lost. It didn't help that I knew laughter increases with intoxication or that Beavis and Butthead share striking similarities with the seventeenth-century Russian folk duo Foma and Yerema. I was alone, with only my hard-earned knowledge of humor to save me.

"So, two hunters from New Jersey were walking in the woods when one accidentally shoots the other. Frantic, the man calls 911. . . . "

〉〉 〈〈

Yes, I told the joke from the LaughLab joke contest that ended the first chapter. But I didn't start with that. I wanted to, though Laura warned me against it. Instead, once I collected myself I told a few warm-up anecdotes about my personal life, then used the humor of the situation—that I was a scientist trying to be funny—to get some audience sympathy. I followed my own advice to relax and be myself. I shared personal stories and let the funniest part of myself shine through. And in the end, I still bombed. So much for the infallibility of science.

The strangest aspect of the show was that I got a few laughs, but they weren't in the places I expected them. It was as though my connection with the audience turned on and off at random points during my performance. And many of my jokes were funnier than those of other performers who got much more applause. That's not just my opinion, it's something several audience members told me afterward. But they shared other things, too. "You held the microphone too low," one lady said as I tried to make a quick exit. "I couldn't hear what you were saying."

Who would have guessed that being unfamiliar with microphones would be such a problem? "I laughed when I heard what you were saying," my friend Jette added, her tone mixed with amusement and pity. "But you sped up a lot too. You're a fast talker, which made you hard to understand. Did you know that?"

Yes, I knew that. I blame six years of living in New England, where it's talk or be talked over. I talk even faster when I'm nervous, which I'm sure made things worse.

Still, I have positive memories of the performance, because there were moments when I felt myself easing into the routine. I simply let my mind go. I wasn't thinking about jokes or the audience, only allowing my unconscious knowledge of humor to express itself naturally. It was a great feeling, although short-lived, and it made me understand why people seek it out. Despite the embarrassment of having performed so badly, then receiving so much advice from friends and strangers afterward, those brief moments made it all worthwhile.

That feeling of being in the moment, which the Hungarian psychologist Mihaly Csikszentmihalyi calls *flow,* is what most athletes and artists strive for, too. Kevin Durant doesn't consciously adjust the arc of his three-point throw just before its release, just as Serena Williams doesn't remind herself to bend her knees as she tosses a ball up for a serve. Our best performance comes when our knowledge, both implicit and explicit, becomes instinct.

Few professional comedians start out successfully because it takes time for humor to become part of who we are, connected to the inner conflicts that define our personalities. When George Carlin started performing, his act was relatively tame, with hardly any cursing or political commentary. He became an icon only after letting loose his contempt for hypocrisy. Richard Pryor didn't attract audiences until he stepped out from Bill Cosby's shadow and tackled race head-on, a subject already prominently on his mind but seldom directly addressed. Steve Martin didn't make it big until he finally accepted that he was the opposite of artists like Carlin and Pryor and embraced his clean-shaven, nonpolitical self by highlighting the farcical aspects of comedy in general.

Though most of us don't aspire to be professional comedians, we can still learn from artists like these by making humor an unconscious part of our lives. When we refer to someone as having a humorous personality, what we mean is that this person sees the ambiguity, confusion, and strife inherent in life and turns them into pleasure. If you really want to be funnier, you can take a seminar—or you can just internalize all you've learned and make it part of a new outlook. By reading this book, you already have the knowledge. All you need to do is use it.

For early organisms on this planet, conflict was simple, involving a single issue: Is something about to kill and eat me? Life soon became more complicated—and so did the human brain. No longer was a puny nerve center enough for keeping us out of trouble. We needed parts for thinking ahead, focusing not just on what might kill us today but also on what might kill us tomorrow. We needed parts for figuring out how to communicate with others and for training future generations not to be killed too. Eventually, millions of years later, we developed parts that began questioning what all these parts were actually for, and why we have so many parts in the first place.

Humor is simply a consequence of having so many parts. It's not wrong we're so complicated, it's just who we are. Some people feel sad most of the time, even though their lives are pretty okay. Some people have to constantly check and recheck locked doors because their anxiety is overwhelming if they don't. These are consequences of owning brains that do so much, and though that might seem like a hassle, consider this—when was the last time a squirrel performed a stand-up comedy routine? A squirrel's brain weighs about 6 grams. With that 6 grams you get a remarkable ability to climb trees and distinguish different kinds of nuts. Multiple that by 250, and you get a whole lot more.

I hope that you've received from this book an appreciation of our complex, modular minds. I also hope you agree that by thinking more deeply about humor we gain a better understanding of how our minds work. Before reading this book you probably knew that surprise is a big part of why jokes are funny. But I doubt you gave much thought to why being surprised by a punch line makes us laugh but being surprised by an intruder doesn't. You probably didn't know that the same chemical responsible for giving drug users a high after snorting cocaine helps us appreciate cartoons and one-liners. Or that simply watching a funny movie lowers stress, improves our immune system response, and even makes us smarter and better problem solvers.

So, the next time you hear a joke that isn't particularly funny, please laugh anyway, knowing that everybody benefits. Not only will you enjoy a happier, healthier life, but others will likely laugh along with you. And it's hard to be in a bad mood while you're laughing.

Oh, and that last joke about the two hunters in the woods, the one I told as part of my comedy routine—it actually got a lot of laughs, more than any other part of my performance. Perhaps it's because I practiced it dozens, maybe hundreds of times. Or maybe it really is the funniest joke in the world. I doubt it, but I recommend practicing it yourself anyway. It never hurts to have a joke or two in your back pocket, in case the occasion ever presents itself.

That's not science, but I stand by the recommendation anyway.

# ACKNOWLEDGMENTS

To Dan and Mary Weems, thank you for teaching me that nothing else is important if you can't laugh.

Thank you to Laura for the constant good humor over twenty years of marriage, especially the last couple in which humor became more than just an outlook. Guess what? Chicken butt!

Special appreciation goes to my friends who attended my comedy performance described in the book's conclusion: Jette Findsen, Brian Goddard, Dave and Roxy Holyoke, Andrew Oliver, and Charlotte Stewart. I'm not sure if I'm happy or sad you got to see the massacre, but at least you know I tried. Thanks also to Magooby's Joke House in Baltimore for not recording the evening.

Thank you to all the scientists who provided interviews and other useful thoughts regarding the book: Salvatore Attardo, Margaret Boden, Jeffrey Burgdorf, Seana Coulson, and Christie Davies. Also thanks to Jenna, who shared her personal stories of gelastic epilepsy. I am especially indebted to Eran Zaidel and James Reggia, who taught me that education, like humor, should continue throughout a lifetime.

Thanks to my agent Ethan Bassoff, who shepherded me through a foreign world, and to the folks at Basic Books for making it all possible.

Much appreciation to Steven Cramer, Leah Hager Cohen, Chris Lynch, and everybody else at Lesley University. You rock.

# Notes

## Introduction

On the frequency of laughter, see Rod Martin and Nicholas Kuiper, "Daily Occurrence of Laughter: Relationships with Age, Gender, and Type A Personality," *Humor: International Journal of Humor Research* 12, no. 4 (1999): 355–384; also Dan Brown and Jennings Bryant, "Humor in the Mass Media," in *Handbook of Humor Research, Volume II: Applied Studies,* eds. Paul McGhee and Jeffrey Goldstein (New York: Springer-Verlag, 1983): 143–172.

On humor and intelligence, see Daniel Howrigan and Kevin MacDonald, "Humor as a Mental Fitness Indicator," *Evolutionary Psychology* 6, no. 4 (2008): 652–666.

On the humor of Albert Camus, see Anne Greenfeld, "Laughter in Camus' *The Stranger, The Fall, and The Renegade*," *Humor: International Journal of Humor Research* 6, no. 4 (1993): 403–414.

On the origins of "Humorology," see Mahadev Apte, "Disciplinary Boundaries in Humorology: An Anthropologist's Ruminations," *Humor: International Journal of Humor Research* 1, no. 1 (1988): 5–25.

## CHAPTER 1: COCAINE, CHOCOLATE, AND MR. BEAN

*Kagera*

On the laughing epidemic in Kagera, see A. Rankin and P. Phillip, "An Epidemic of Laughing in the Bukoba District of Tanganyika," *Central African Journal of Medicine* 9 (1963): 167–170; also Christian Hempelmann, "The Laughter of the 1962 Tanganyika Laughter Epidemic," *Humor: International Journal of Humor Research* 20, no. 1 (2007): 49–71.

*What Is Humor?*

For the interview with Conchesta, as well as an informative review of laughter in general, I recommend *RadioLab*'s excellent podcast titled "Laughter."

On chimpanzee laughter, see Frans de Waal, *Chimpanzee Politics: Power and Sex Among Apes* (New York: Harper & Row, 1982).

On dog laughter, see Patricia Simonet, Donna Versteeg, and Dan Storie, "Dog-Laughter: Recorded Playback Reduces Stress-Related Behavior in Shelter Dogs," in *Proceedings of the 7th International Conference on Environmental Enrichment* (New York, 2005).

*The Elusive Concept of Mirth*

On the Mirth Response Test, see Jacob Levine and Robert Abelson, "Humor as a Disturbing Stimulus," in *Motivation in Humor,* ed. Jacob Levine (New Brunswick, NJ: Transaction Publishers, 1969), pp. 38–48.

On humorous cartoons and the dopamine reward circuit, see Dean Mobbs, Michael Greicius, Eiman Abdel-Azim, Vinod Menon, and Allan Reiss, "Humor Modulates the Mesolimbic Reward Centers," *Neuron* 40 (2003): 1041–1048.

On musical chills, see Anne Blood and Robert Zatorre, "Intensely Pleasurable Responses to Music Correlate with Activity in Brain Regions Implicated in Reward and Emotion," *Proceedings of the National Academy of Sciences* 98, no. 20 (2001): 11818–11823.

On Mr. Bean and dopamine rewards, see Masao Iwase, Yasuomi Ouchi, Hiroyuki Okada, Chihiro Yokoyama, Shuji Nobezawa, Etsuji Yoshikawa, Hideo Tsukada, Masaki Takeda, Ko Yamashita, Masatoshi Takeda, Kouzi Yamaguti, Hirohiko Kuratsune, Akira Shimizu, and

Yasuyoshi Watanabe, "Neural Substrates of Human Facial Expression of Pleasant Emotion Induced by Comic Films: A PET Study," *NeuroImage* 17 (2002): 758–768.

On rat vocalizations, see Jeffrey Burgdorf, Paul Wood, Roger Kroes, Joseph Moskal, and Jaak Panksepp, "Neurobiology of 50-kHz Ultrasonic Vocalizations in Rats: Electrode Mapping, Lesion, and Pharmacology Studies," *Behavioral Brain Research* 182 (2007): 274–283; also Jaak Panksepp and Jeff Burgdorf, "Laughing Rats and the Evolutionary Antecedents of Human Joy?" *Physiology and Behavior* 79 (2003): 533–547; also Jaak Panksepp and Jeffrey Burgdorf, "Laughing Rats? Playful Tickling Arouses High-Frequency Ultrasonic Chirping in Young Rodents," in *Toward a Science of Consciousness III: The Third Tucson Discussions and Debates,* eds. Stuart Hameroff, Alfred Kaszniak, and David Chalmers (Cambridge, MA: MIT Press, 1999).

### The Funniest Joke in the World

On the LaughLab experiment, see Richard Wiseman's *Quirkology: How We Discover the Big Truths in Small Things* (New York: Basic Books, 2008).

## CHAPTER 2: THE KICK OF THE DISCOVERY

On explanation as a major drive, see Alison Gopnik, "Explanation as Orgasm," *Minds and Machines* 8 (1998): 101–118.

On word triads and positive facial reactions, see Sascha Topolinski, Katja Likowski, Peter Weyers, and Fritz Strack, "The Face of Fluency: Semantic Coherence Automatically Elicits a Specific Pattern of Facial Muscle Reactions," *Cognition and Emotion* 23, no. 2 (2009): 260–271.

On insight and positive mood, see Karuna Subramaniam, John Kounios, Todd Parrish, and Mark Jung-Beeman, "A Brain Mechanism for Facilitation of Insight by Positive Affect," *Journal of Cognitive Neuroscience* 21, no. 3 (2008): 415–432. For more on semantic associates, see Edward Bowden and Mark Jung-Beeman, "Normative Data for 144 Compound Remote Associate Problems," *Behavior Research Methods, Instruments, and Computers* 35, no. 4 (2003): 634–639.

### Constructing and the Anterior Cingulate

On emotion and the Stroop task, see Julius Kuhl and Miguel Kazén, "Volitional Facilitation of Difficult Intentions: Joint Activation of

Intention Memory and Positive Affect Removes Stroop Interference," *Journal of Experimental Psychology: General* 128, no. 3 (1999): 382–399.

On weight judging and laughter, see Göran Nerhardt, "Humor and Inclination to Laugh: Emotional Reactions to Stimuli of Different Divergence from a Range of Expectancy," *Scandinavian Journal of Psychology* 11 (1970): 185–195; also Lambert Deckers, "On the Validity of a Weight-Judging Paradigm for the Study of Humor," *Humor: International Journal of Humor Research* 6, no. 1 (1993): 43–56.

### Reckoning in a Confusing World

On confidence and insight tasks, see Janet Metcalfe, "Premonitions of Insight Predict Impending Error," *Journal of Experimental Psychology: Learning, Memory, and Cognition* 12, no. 4 (1986): 623–634.

On surprise and pleasure, see Craig Smith and Phoebe Ellsworth, "Patterns of Cognitive Appraisal in Emotion," *Journal of Personality and Social Psychology* 48, no. 4 (1985): 813–838.

On the pleasantness of surprise in music and architecture, see Rudolf Arnheim, *The Dynamics of Architectural Form* (Los Angeles: University of California Press, 1977).

On brain activation during cartoon viewing, see Karli Watson, Benjamin Matthews, and John Allman, "Brain Activation During Sight Gags and Language-Dependent Humor," *Cerebral Cortex* 17 (2007): 314–324.

### Resolving with Scripts

On scripts, see Salvatore Attardo, Christian Hempelmann, and Sara Di Maio, "Script Oppositions and Logical Mechanisms: Modeling Incongruities and Their Resolutions," *Humor: International Journal of Humor Research* 15, no. 1 (2002): 3–46.

On the General Theory of Verbal Humor, see Salvatore Attardo and Victor Raskin, "Script Theory Revisited: Joke Similarity and Joke Representation Model," *Humor: International Journal of Humor Research* 4, no. 3/4 (1991): 293–347.

On background incongruities in jokes, see Andrea Samson and Christian Hempelmann, "Humor with Background Incongruity: Does More Required Suspension of Disbelief Affect Humor Perception?" *Humor: International Journal of Humor Research* 24, no. 2 (2011): 167–185.

On EEG responses to jokes and punch lines, see Peter Derks, Lynn Gil-likin, Debbie Bartolome-Rull, and Edward Bogart, "Laughter and Electroencephalographic Activity," *Humor: International Journal of Humor Research* 10, no. 3 (1997): 285–300.

*Beyond the Stages*

On ambiguity in headlines, see Chiara Bucaria, "Lexical and Syntactic Ambiguity as a Source of Humor: The Case of Newspaper Headlines," *Humor: International Journal of Humor Research* 17, no. 3 (2004): 279–309.

On political orientation and the anterior cingulate, see Ryota Kanai, Tom Feilden, Colin Firth, and Geraint Rees, "Political Orientations Are Correlated with Brain Structure in Young Adults," *Current Biology* 21 (2011): 677–680.

On brain activity and religious belief, see Michael Inzlicht and Alexa Tullett, "Reflecting on God: Religious Primes Can Reduce Neurophysiological Response to Errors," *Psychological Science* 21, no. 8 (2010): 1184–1190; also, interesting background in James Austin, *Zen and the Brain* (New York: MIT Press, 1998).

## CHAPTER 3: STOPOVER AT THE EMPIRE STATE BUILDING

*Humor Gets a Bad Rap*

On humor and the Bible, see John Morreall, "Comic Vices and Comic Virtues," *Humor: International Journal of Humor Research* 23, no. 1 (2010): 1–26; also John Morreall, "Philosophy and Religion," in *The Primer of Humor Research*, ed. Victor Raskin (New York: Mouton de Gruyter, 2009), pp. 211–228; also Jon Roeckelein, *The Psychology of Humor: A Reference Guide and Annotated Bibliography* (Westport, CT: Greenwood Press, 2002).

On joke latencies following disasters, see Bill Ellis, "A Model for Collecting and Interpreting World Trade Center Disaster Jokes," *New Directions in Folklore* 5 (2001): 1–9.

On insult humor, see Christie Davies's "Undertaking the Comparative Study of Humor," in *The Primer of Humor Research*, ed. Victor Raskin (New York: Mouton de Gruyter, 2009), pp. 162–175; also Christie Davies, *Ethnic Humor Around the World: A Comparative Analysis* (Indianapolis: Indiana University Press, 1990); also *Mirth of Nations* (New

Brunswick, NJ: Transaction Publishers, 2002). The quotations in the text are from personal interviews.

On jokes about the handicapped, see Herbert Lefcourt and Rod Martin, *Humor and Life Stress: Antidote to Adversity* (New York: Springer-Verlag, 1986).

On humor and the recovery process, see Dacher Keltner and George Bonanno, "A Study of Laughter and Dissociation: Distinct Correlates of Laughter and Smiling During Bereavement," *Journal of Personality and Social Psychology* 73, no. 4 (1997): 687–702; also Charles Carver, Christina Pozo, Suzanne Harris, Victoria Noriega, Michael Scheier, David Robinson, Alfred Ketcham, Frederick Moffat, and Kimberly Clark, "How Coping Mediates the Effect of Optimism on Distress: A Study of Women with Early Stage Breast Cancer," *Journal of Personality and Social Psychology* 65, no. 2 (1993): 375–390.

On the role of cruelty in humor, see Thomas Herzog and Beverly Bush, "The Prediction of Preference for Sick Humor," *Humor: International Journal of Humor Research* 7, no. 4 (1994): 323–340; also Thomas Herzog and Joseph Karafa, "Preferences for Sick Versus Nonsick Humor," *Humor: International Journal of Humor Research* 11, no. 3 (1998): 291–312; also Thomas Herzog and Maegan Anderson, "Joke Cruelty, Emotional Responsiveness, and Joke Appreciation," *Humor: International Journal of Humor Research* 13, no. 3 (2000): 333–351.

On humor in the media following disasters, see Giselinde Kuipers, "Where Was King Kong When We Needed Him? Public Discourse, Digital Disaster Jokes, and the Functions of Laughter after 9/11," *Journal of American Culture* 28, no. 1 (2005): 70–84.

*Scary Movies and Relief*

On emotional experience during horror movies, see Eduardo Andrade and Joel Cohen, "On the Consumption of Negative Feelings," *Journal of Consumer Research* 34 (2007): 283–300.

On Great Humor, see Hans Vejleskov, "A Distinction Between Small Humor and Great Humor and Its Relevance to the Study of Children's Humor," *Humor: International Journal of Humor Research* 14, no. 4 (2001): 323–338.

On prisoner-of-war humor, including the story of Gerald Santo Venanzi, see Linda Henman, "Humor as a Coping Mechanism:

Lessons from POWs," *Humor: International Journal of Humor Research* 14, no. 1 (2001): 83–94.

*Jokes with a Target*

For a general overview on spindle cells, see John Allman, Atiya Hakeem, and Karli Watson, "The Phylogenetic Specializations in the Human Brain," *The Neuroscientist* 8, no. 4 (2002): 335–346; also Karli Watson, T. K. Jones, and John Allman, "Dendritic Architecture of the Von Economo Neurons," *Neuroscience* 141 (2006): 1107–1112.

On the Emotional Stroop task, see John Allman, Atiya Hakeem, Joseph Erwin, Esther Nimchinsky, and Patrick Hof, "The Anterior Cingulate: The Evolution of an Interface Between Emotion and Cognition," *Annals of the New York Academy of Sciences* 935 (2001): 107–117.

On David Levy jokes, see Hagar Salamon, "The Ambivalence over the Levantinization of Israel: David Levi Jokes," *Humor: International Journal of Humor Research* 20, no. 4 (2007): 415–442.

On elephant jokes and latent racism, see Roger Abrahams and Alan Dundes, "On Elephantasy and Elephanticide," *Psychoanalysis Review* 56 (1969): 225–241.

On lawyer jokes, see Christie Davies, "American Jokes About Lawyers," *Humor: International Journal of Humor Research* 21, no. 4 (2008): 369–386; also Marc Galanter, "The Great American Lawyer Joke Explosion," *Humor: International Journal of Humor Research* 21, no. 4 (2008): 387–413.

On the Dyak tribes of Borneo, see V. I. Zelvys, "Obscene Humor: What the Hell?" *Humor: International Journal of Humor Research* 3, no. 3 (1990): 323–332.

## CHAPTER 4: SPECIALIZATION IS FOR INSECTS

*A.K.*

On patient A.K., see Itzhak Fried, Charles Wilson, Katherine MacDonald, and Eric Behnke, "Electric Current Stimulates Laughter," *Nature* 391 (1998): 650.

On gelastic epilepsy, see R. Garg, S. Misra, and R. Verma, "Pathological Laughter as Heralding Manifestation of Left Middle Cerebral Artery Territory Infarct: Case Report and Review of the Literature," *Neurology India* 48 (2000): 388–390; also Mario Mendez, Tomoko

Nakawatase, and Charles Brown, "Involuntary Laughter and In-appropriate Hilarity," *Journal of Neuropsychiatry and Clinical Neurosciences* 11, no. 2 (1999): 253–258.

## States and Traits

On Peter Derks's humor formula, see Antony Chapman and Hugh Foot, *Humor and Laughter: Theory, Research, and Applications* (New Brunswick, NJ: Transaction Publishers, 1996).

On humor and religiosity, see Vassilis Saroglou, "Being Religious Implies Being Different in Humour: Evidence from Self- and Peer Ratings," *Mental Health, Religion, and Culture* 7, no. 3 (2004): 255–267.

On the personality characteristics of cartoonists, see Paul Pearson, "Personality Characteristics of Cartoonists," *Personality and Individual Differences* 4, no. 2 (1983): 227–228.

On gender differences for Eysenck's personality traits, see R. Lynn and T. Martin, "Gender Differences in Extraversion, Neuroticism, and Psychoticism in 37 Nations," *Journal of Social Psychology* 137, no. 3 (1997): 369–373.

On the personality characteristics of creative people, see Giles Burch, Christos Pavelis, David Hemsley, and Philip Corr, "Schizotypy and Creativity in Visual Artists," *British Journal of Psychology* 97 (2006): 177–190; also Gregory Feist, "A Meta-Analysis of Personality in Scientific and Artistic Creativity," *Personality and Social Psychology Review* 2, no. 4 (1998): 290–309; also Karl Gotz and Karin Gotz, "Personality Characteristics of Successful Artists," *Perceptual and Motor Skills* 49 (1979): 919–924; also Cary Cooper and Geoffrey Wills, "Popular Musicians Under Pressure," *Psychology of Music* 17, no. 1 (1989): 22–36.

On Willibald Ruch's large-scale study of sense of humor and personality characteristics, see Gabrielle Köhler and Willibald Ruch, "Sources of Variance in Current Sense of Humor Inventories: How Much Substance, How Much Method Variance?" *Humor: International Journal of Humor Research* 9, no. 3/4 (1996): 363–397.

On sensation-seekers and absurd humor, see Andrea Samson, Christian Hempelmann, Oswald Huber, and Stefan Zysset, "Neural Substrates of Incongruity-Resolution and Nonsense Humor," *Neuropsychologia* 47 (2009): 1023–1033.

On humor and environmentalism, see Herbert Lefcourt, "Perspective-Taking Humor and Authoritarianism as Predictors of Anthropocentrism," *Humor: International Journal of Humor Research* 9, no. 1 (1996): 57–71.

On humor and Type A personalities, see Rod Martin and Nicholas Kuiper, "Daily Occurrence of Laughter: Relationships with Age, Gender, and Type A Personality," *Humor: International Journal of Humor Research* 12, no. 4 (1999): 355–384.

On humor and anality, see Richard O'Neill, Roger Greenberg, and Seymour Fisher, "Humor and Anality," *Humor: International Journal of Humor Research* 5, no. 3 (1992): 283–291.

## The Fairer Sex

On Robin Lakoff's take on feminism and humor, see her book *Language and Woman's Place* (New York: Oxford University Press, 2004).

On laughter in natural settings, see Robert Provine, *Laughter: A Scientific Investigation* (New York: Penguin, 2001).

On sex differences in brain activation during jokes, see Eiman Azim, Dean Mobbs, Booil Jo, Vinod Menon, and Allan Reiss, "Sex Differences in Brain Activation Elicited by Humor," *Proceedings of the National Academy of Sciences* 102, no. 45 (2005): 16496–16501.

On cartoons from *Playboy* versus *The New Yorker,* see Peter Derks, "Category and Ratio Scaling of Sexual and Innocent Cartoons," *Humor: International Journal of Humor Research* 5, no. 4 (1992): 319–329.

On the consequences of sexist humor, see Thomas Ford, Christie Boxer, Jacob Armstrong, and Jessica Edel, "More Than Just a Joke: The Prejudice-Releasing Function of Sexist Humor," *Personality and Social Psychology Bulletin* 34, no. 2 (2008): 159–170.

## Specialization Is for Insects

On object permanence in animals, see Francois Doré, "Object Permanence in Adult Cats (*Felis Catus*)," *Journal of Comparative Psychology* 100, no. 4 (1986): 340–347; also Holly Miller, Cassie Gipson, Aubrey Vaughn, Rebecca Rayburn-Reeves, and Thomas Zentall, "Object Permanence in Dogs: Invisible Displacement in a Rotation Task," *Psychonomic Bulletin and Review* 16, no. 1 (2009): 150–155; also Almut Hoffmann, Vanessa

Rüttler, and Andreas Nieder, "Ontogeny of Object Permanence and Object Tracking in the Carrion Crow, *Corvus Corone*," *Animal Behavior* 82 (2011): 359–367.

On children's learning of irony and sarcasm, see Amy Demorest, Christine Meyer, Erin Phelps, Howard Gardner, and Ellen Winner, "Words Speak Louder Than Actions: Understanding Deliberately False Remarks," *Child Development* 55 (1984): 1527–1534; also Carol Capelli, Noreen Nakagawa, and Cory Madden, "How Children Understand Sarcasm: The Role of Context and Intonation," *Child Development* 61 (1990): 1824–1841.

On humor and conservatism, see Willibald Ruch, Paul McGhee, and Franz-Josef Hehl, "Age Differences in the Enjoyment of Incongruity-Resolution and Nonsense Humor During Adulthood," *Personality and Aging* 5, no. 3 (1990): 348–355.

## CHAPTER 5: OUR COMPUTER OVERLORDS

On Watson's victory, see Stephen Baker, *Final Jeopardy: Man Versus Machine and the Quest to Know Everything* (New York: Houghton Mifflin Harcourt, 2011). On Watson's design, see the white paper released by IBM titled "Watson—A System Designed for Answers," which can easily be found using online search.

### Pattern Detection and Hypothesis Generation

To see The Joking Computer, you can visit the public website at http://www.abdn.ac.uk/jokingcomputer; for an excellent website that chooses jokes specifically matching your own sense of humor based on a filtering algorithm, see http://eigentaste.berkeley.edu.

On humor, computers, and creativity, see almost anything by Margaret Boden, including *The Creative Mind: Myths and Mechanisms* (New York: Routledge, 2004); also "Creativity and Artificial Intelligence," *Artificial Intelligence* 103 (1998): 347–356; also "Creativity and Computers," *Current Science* 64, no. 6 (1993): 419–433. The quotations in the text are from personal interviews.

On JAPE, see Kim Binstead and Graeme Ritchie, "An Implemented Model of Punning Riddles," in *Proceedings of the Twelfth National Conference on Artificial Intelligence* (Menlo Park, CA: American Association for Artificial Intelligence, 1994).

On Hahacronym, see Oliviero Stock and Carlo Strapparava, "Haha-cronym: A Computational Humor System," in *Proceedings of the ACL Interactive Poster and Demonstration Sessions* (Ann Arbor, MI: Association for Computational Linguistics, 2005); also Oliviero Stock and Carlo Strapparava, "Hahacronym: Humorous Agents for Humorous Acronyms," *Humor: International Journal of Humor Research* 16, no. 3 (2003): 297–314.

On DEviaNT, see Chloé Kiddon and Yuriy Brun, "That's What She Said: Double Entendre Identification," in *Proceedings of the 49th Annual Meeting of the Association for Computational Linguistics* (Portland, OR: Association for Computational Linguistics, 2011).

On the University of North Texas's one-liner computer program, see Rada Mihalcea and Carlo Strapparava, "Making Computers Laugh: Investigations in Automatic Humor Recognition," in *Proceedings of the Joint Conference on Human Language Technology/Empirical Methods in Natural Language Processing* (Vancouver, Canada, 2005); also Rada Mihalcea and Carlo Strapparava, "Learning to Laugh (Automatically): Computational Models for Humor Recognition," *Computational Intelligence* 22, no. 2 (2006): 126–142.

On cloze probability and humor, see Rachel Giora, "Optimal Innovation and Pleasure," in *Proceedings of the Twentieth Workshop on Language Technology* (Trento, Italy, 2002).

On humor and N400 effects, see Seana Coulson and Marta Kutas, "Getting It: Human Event-Related Brain Response to Jokes in Good and Poor Comprehenders," *Neuroscience Letters* 316 (2001): 71–74.

On semantic priming and humor, see Jyotsna Vaid, Rachel Hull, Roberto Heredia, David Gerkens, and Francisco Martinez, "Getting the Joke: The Time Course of Meaning Activation in Verbal Humor," *Journal of Pragmatics* 35 (2003): 1431–1449.

*Transformational Creativity*

On the neuroscience of creativity, see Arne Dietrich and Riam Kanso, "A Review of EEG, ERP, and Neuroimaging Studies of Creativity and Insight," *Psychological Bulletin* 136, no. 5 (2010): 822–848; also Hikaru Takeuchi, Yasuyuki Taki, Hiroshi Hashizume, Yuko Sassa, Tomomi Nagase, Rui Nouchi, and Ryuta Kawashima, "The Association

Between Resting Functional Connectivity and Creativity," *Cerebral Cortex* 22, no. 12 (2012): 1–9.

On Gaiku, see Yael Netzer, David Gabay, Yoav Goldberg, and Michael Elhadad, "Gaiku: Generating Haiku with Word Association Norms," in *NAACL Workshop on Computational Approaches to Linguistic Creativity* (Boulder, CO, 2009).

On computer attempts to model music, painting, and other arts, see Boden's *The Creative Mind: Myths and Mechanisms;* also Paul Hodgson, "Modeling Cognition in Creative Musical Improvisation," unpublished doctoral thesis, University of Sussex Department of Informatics; also H. Koning and J. Eizenberg, "The Language of the Prairie: Frank Lloyd Wright's Prairie Houses," *Environmental Planning B* 8 (1981): 295–323; also James Meehan, "The Metanovel: Writing Stories by Computer," unpublished doctoral thesis, Yale University Department of Computer Science; also Patrick McNally and Kristian Hammond, "Picasso, Pato, and Perro: Reconciling Procedure with Creativity," in *Proceedings of the International Conference on Computational Creativity* (Mexico City, Mexico, 2011); also Harold Cohen, *On the Modeling of Creative Behavior* (Santa Monica, CA: Rand Corporation Technical Paper, 1981).

*Keeping Salt Out*

On measurement of creativity, see Mary Lou Maher, "Evaluating Creativity in Humans, Computers, and Collectively Intelligent Systems," in *Proceedings of the First DESIRE Network Conference on Creativity and Innovation in Design* (Lancaster, England, 2010); also Graeme Ritchie, "Some Empirical Criteria for Attributing Creativity to a Computer Program," *Minds and Machines* 17 (2007): 67–99.

On The Automatic Mathematician, see G. Ritchie and F. Hanna, "Automatic Mathematician: A Case Study in AI Methodology," *Artificial Intelligence* 23 (1984): 249–258.

On the Chinese Room Thought Experiment, see John Searle, "Minds, Brains, and Programs," *Behavioral and Brain Sciences* 3, no. 3 (1980): 417–457.

CHAPTER 6: THE BILL COSBY EFFECT

On humor and illness, see Norman Cousins, *Anatomy of an Illness as Perceived by the Patient* (New York: W. W. Norton, 1979).

## The Doctor Inside

On laughter as exercise, see M. Buchowski, K. Majchrzak, K. Blom-
quist, K. Chen, D. Byrne, and J. Bachorowski, "Energy Expenditure
of Genuine Laughter," *International Journal of Obesity* 31 (2007):
131–137.

On laughter and blood pressure, see William Fry and William Savin,
"Mirthful Laughter and Blood Pressure," *Humor: International Journal of Humor Research* 1, no. 1 (1988): 49–62; also Jun Sugawara,
Takashi Tarumi, and Hirofumi Tanaka, "Effect of Mirthful Laughter on Vascular Function," *American Journal of Cardiology* 106, no. 6
(2010): 856–859.

On Michael Miller's vasoreactivity studies, see Michael Miller and William Fry, "The Effect of Mirthful Laughter on the Human Cardiovascular System," *Medical Hypotheses* 73, no. 5 (2009): 636–643; also see
accounts of Michael Miller's presentation to the *Scientific Session of
the American College of Cardiology* (Orlando, FL, 2005).

On laughter and diabetes, see Takashi Hayashi, Osamu Urayama, Miyo
Hori, Shigeko Sakamoto, Uddin Mohammad Nasir, Shizuko Iwanaga,
Keiko Hayashi, Fumiaki Suzuki, Koichi Kawai, and Kazuo Murakami,
"Laughter Modulates Prorenin Receptor Gene Expression in Patients
With Type 2 Diabetes," *Journal of Psychonomic Research* 62 (2007):
703–706; also Keiko Hayashi, Takashi Hayashi, Shizuko Iwanaga,
Koichi Kawai, Hitoshi Ishii, Shin'ichi Shoji, and Kanuo Murakami,
"Laughter Lowered the Increase in Postprandial Blood Glucose," *Dia-
betes Care* 26, no. 5 (2003): 1651–1652.

For a review on humor and diseases such as arthritis and dermatitis,
see Paul McGhee, *Humor: The Lighter Path to Resilience and Health*
(Bloomington, IN: AuthorHouse, 2010).

On humor and the immune system, see Herbert Lefcourt, Karina
Davidson-Katz, and Karen Kueneman, "Humor and Immune-System
Functioning," *Humor: International Journal of Humor Research* 3, no.
3 (1990): 305–321; also Arthur Stone, Donald Cox, Heiddis Valdi-
marsdottir, Lina Jandorf, and John Neale, "Evidence That Secretory
IgA Antibody Is Associated with Daily Mood," *Journal of Personality
and Social Psychology* 52, no. 5 (1987): 988–993; also Mary Bennett,
Janice Zeller, Lisa Rosenberg, and Judith McCann, "The Effect of
Mirthful Laughter on Stress and Natural Killer Cell Activity," *Alternative Therapies* 9, no. 2 (2003): 38–44.

On the Norwegian health study, see Sven Svebak, Rod Martin, and Jostein Holmen, "The Prevalence of Sense of Humor in a Large, Unselected Country Population in Norway: Relations with Age, Sex, and Some Health Indicators," *Humor: International Journal of Humor Research* 17, no. 1/2 (2004): 121–134.

On personality and longevity, see Howard Friedman, Joan Tucker, Carol Tomlinson-Keasey, Joseph Schwartz, Deborah Wingard, and Michael Criqui, "Does Childhood Personality Predict Longevity?" *Journal of Personality and Social Psychology* 65, no. 1 (1993): 176–185.

On humor and heart-unhealthy traits, see Paavo Kerkkänen, Nicholas Kuiper, and Rod Martin, "Sense of Humor, Physical Health, and Well-Being at Work: A Three-Year Longitudinal Study of Finnish Police Officers," *Humor: International Journal of Humor Research* 17, no. 1/2 (2004): 21–35.

On neuroticism and longevity, see Benjamin Lahey, "Public Health Significance of Neuroticism," *American Psychologist* 64, no. 4 (2009): 241–256.

### The Bill Cosby Effect

On humor and recovery of hospital patients, see James Rotton and Mark Shats, "Effects of State Humor, Expectancies, and Choice on Postsurgical Mood and Self-Medication: A Field Experiment," *Journal of Applied Social Psychology* 26, no. 20 (1996): 1775–1794.

On humor, pain tolerance, and the cold pressor test, see Matisyohu Weisenberg, Inbal Tepper, and Joseph Schwarzwald, "Humor as a Cognitive Technique for Increasing Pain Tolerance," *Pain* 63 (1995): 207–212.

On the benefits of watching sitcoms like *Friends,* compared to sitting and resting, see Attila Szabo, Sarah Ainsworth, and Philippa Danks, "Experimental Comparison of the Psychological Benefits of Aerobic Exercise, Humor, and Music," *Humor: International Journal of Humor Research* 18, no. 3 (2005): 235–246.

On humor styles and health, see Paul Frewen, Jaylene Brinker, Rod Martin, and David Dozois, "Humor Styles and Personality—Vulnerability to Depression," *Humor: International Journal of Humor Research* 21, no. 2 (2008): 179–195; also Vassilis Saroglou and Lydwine Anciaux, "Liking Sick Humor: Coping Styles and Religion as Predictors," *Humor: International Journal of Humor Research* 17, no. 3 (2004): 257–277; also Nicholas Kuiper and Rod Martin, "Humor and Self-Concept," *Humor: International Journal of Humor Research* 6, no. 3 (1993): 251–270; also Nicholas Kuiper, Melissa Grimshaw, Catherine Leite, and Gillian Kirsh, "Humor Is Not Always the Best Medicine: Specific Components of Sense of Humor and Psychological Well-Being," *Humor: International Journal of Humor Research* 17, no. 1/2 (2004): 135–168.

On the moderator hypothesis of humor (discussed on page 147), see Arthur Nezu, Christine Nezu, and Sonia Blissett, "Sense of Humor as a Moderator of the Relation Between Stressful Events and Psychological Distress: A Prospective Analysis," *Journal of Personality and Social Psychology* 54, no. 3 (1988): 520–525.

On humor and the movie *Faces of Death,* see Arnie Cann, Lawrence Calhoun, and Jamey Nance, "Exposure to Humor Before and After an Unpleasant Stimulus: Humor as a Preventative or a Cure," *Humor: International Journal of Humor Research* 13, no. 2 (2000): 177–191.

On humor and positive outlook, see Millicent Abel, "Humor, Stress, and Coping Strategies," *Humor: International Journal of Humor Research* 15, no. 4 (2002): 365–381; also N. Kuiper, R. Martin, and K. Dance, "Sense of Humor and Enhanced Quality of Life," *Personality and Individual Differences* 13, no. 12 (1992): 1273–1283.

For examples of humor in hospitals, see John Morreall, "Applications of Humor: Health, the Workplace, and Education," in *The Primer of Humor Research,* ed. Victor Raskin (New York: Mouton de Gruyter, 2009); also Paul McGhee's *Humor: The Lighter Path to Resilience and Health* (New York: AuthorHouse, 2010).

## CHAPTER 7: HUMOR DANCES

### Humor and Dancing

On similarities between humor and jazz, see Kendall Walton, "Understanding Humor and Understanding Music," *The Journal of Musicology* 11, no. 1 (1993): 32–44; also Frank Salamone, "Close Enough for Jazz: Humor and Jazz Reality," *Humor: International Journal of Humor Research* 1, no. 4 (1988): 371–388.

On comedic timing, see Salvatore Attardo and Lucy Pickering, "Timing in the Performance of Jokes," *Humor: International Journal of Humor Research* 24, no. 2 (2011): 233–250.

On paratones, see Lucy Pickering, Marcella Corduas, Jodi Eisterhold, Brenna Seifried, Alyson Eggleston, and Salvatore Attardo, "Prosodic Markers of Saliency in Humorous Narratives," *Discourse Processes* 46 (2009): 517–540.

On jab lines, see Villy Tsakona, "Jab Lines in Narrative Jokes," *Humor: International Journal of Humor Research* 16, no. 3 (2003): 315–329.

For a review on Paul Grice and his rules of communication, see Daniel Perlmutter, "On Incongruities and Logical Inconsistencies in Humor: The Delicate Balance," *Humor: International Journal of Humor Research* 15, no. 2 (2002): 155–168; also Salvatore Attardo's *Linguistic Theories of Humor* (New York: Mouton de Gruyter, 1994).

On the uniqueness of irony, see Salvatore Attardo, Jodi Eisterhold, Jennifer Hay, and Isabella Poggi, "Multimodal Markers of Irony and Sarcasm," *Humor: International Journal of Humor Research* 16, no. 2 (2003): 243–260.

*Peer Pressure*

On experimenters' influencing of participant humor ratings, see Willibald Ruch, "State and Trait Cheerfulness and the Induction of Exhilaration: A FACS Study," *European Psychologist* 2, no. 4 (1997): 328–341.

On shared laughter, see Howard Pollio and Charles Swanson, "A Behavioral and Phenomenological Analysis of Audience Reactions to Comic Performance," *Humor: International Journal of Humor Research* 8, no. 1 (1995): 5–28; also Jonathan Freedman and Deborah Perlick, "Crowding, Contagion, and Laughter," *Journal of Experimental Social Psychology* 15 (1979): 295–303; also Jennifer Butcher and Cynthia Whissell, "Laughter as a Function of Audience Size, Sex of the Audience, and Segments of the Short Film 'Duck Soup,'" *Perceptual and Motor Skills* 59 (1984): 949–950; also Alan Fridlund, "Sociality of Solitary Smiling: Potentiation by an Implicit Audience," *Journal of Personality and Social Psychology* 60, no. 2 (1991): 229–240; also T. Nosanchuk and Jack Lightstone, "Canned Laughter and Public and Private Conformity," *Journal of Personality and Social Psychology* 29, no. 1 (1974): 153–156; also Richard David Young and Margaret Frye, "Some Are Laughing, Some Are Not—Why?" *Psychological Reports* 18 (1966): 747–754.

On experimental manipulations of humor, see David Wimer and Bernard Beins, "Expectations and Perceived Humor," *Humor: International Journal of Humor Research* 21, no. 3 (2008): 347–363; also James Olson and Neal Roese, "The Perceived Funniness of Humorous Stimuli," *Personality and Social Psychology Bulletin* 21, no. 9 (1995): 908–913; also Timothy Lawson, Brian Downing, and Hank Cetola, "An Attributional Explanation for the Effect of Audience Laughter on Perceived Funniness," *Basic and Applied Social Psychology* 20, no. 4 (1998): 243–249.

Lastly, for any old high school friends who are wondering (regarding the section's final metaphor): no, I wasn't related to my prom date. Susan, wherever you are now, I hope all is well.

## Two Brains, One Mind

I changed Linda's name out of respect for her privacy. In academic literature she is known as patient N.G. I did the same for Philip, who is known as patient L.B.

On split brains and hemispheric laterality, see Eran Zaidel and Marco Iacoboni, *The Parallel Brain: The Cognitive Neuroscience of the Corpus Callosum* (Cambridge, MA: MIT Press, 2003). The quotations in the text are from personal interviews. On the commissurotomy process, see Joseph Bogen and Philip Vogel, "Neurologic Status in the Long Term Following Complete Cerebral Commissurotomy," in F. Michel and B. Schott, *Les Syndromes de Disconnexion Calleuse Chez l'Homme* (Hôpital Lyon, 1974).

On humor loss in right-hemisphere-damaged patients, see Hiram Brownell, Dee Michel, John Powelson, and Howard Gardner, "Surprise But Not Coherence: Sensitivity to Verbal Humor in Right-Hemisphere Patients," *Brain and Language* 18 (1983): 20–27.

On general personality differences between the hemispheres, see Fredric Schiffer, Eran Zaidel, Joseph Bogen, and Scott Chasan-Taber, "Different Psychological Status in the Two Hemispheres of Two Split-Brain Patients," *Neuropsychiatry, Neuropsychology, and Behavioral Neurology* 11, no. 3 (1998): 151–156; also the talk presented by Vilayanur Ramachandran to the 2006 Beyond Belief Conference at the Salk Institute for Biological Studies in La Jolla, California, available freely on YouTube.

On right-hemisphere importance for insight and poetry, see Edward Bowden, Mark Jung-Beeman, Jessica Fleck, and John Kounios, "New Approaches to Demystifying Insight," *Trends in Cognitive Sciences* 9, no. 7 (2005): 322–328; also Edward Bowden and Mark Jung-Beeman, "Aha! Insight Experience Correlates with Solution Activation in the Right Hemisphere," *Psychonomic Bulletin and Review* 10, no. 3 (2003): 730–737; also Edward Bowden and Mark Jung-Beeman, "Getting the Right Idea: Semantic Activation in the Right Hemisphere May Help Solve Insight Problems," *Psychological Science* 9, no. 6 (1988): 435–440; also M. Faust and N. Mashal, "The Role of the Right Cerebral Hemisphere in Processing Novel Metaphoric Expressions Taken from Poetry: A Divided Visual Field Study," *Neuropsychologia* 45 (2007): 860–870.

## Funny Relationships

On humor and mate selection, see Jane Smith, Ann Waldorf, and David Trembath, "Single, White Male Looking for Thin, Very Attractive . . . ," *Sex Roles* 23, no. 11 (1990): 675–685; also Hal Daniel, Kevin O'Brien, Robert McCabe, and Valerie Quinter, "Values in Mate Selection: A 1984 Campus Study," *College Student Journal* 15 (1986): 44–50; also Bojan Todosijević, Snežana Ljubinković, and Aleksandra Arančić, "Mate Selection Criteria: A Trait Desirability Assessment Study of Sex Differences in Serbia," *Evolutionary Psychology* 1 (2003): 116–126; also Lester Hewitt, "Student Perceptions of Traits Desired in Themselves as Dating and Marriage Partners," *Marriage and Family Living* 20, no. 4 (1958): 344–349; also Richard Lippa, "The Preferred Traits of Mates in a Cross-National Study of Heterosexual and Homosexual Men and Women: An Examination of Biological and Cultural Influences," *Archives of Sexual Behavior* 36 (2007): 193–208.

On gender differences in humor production and appreciation, see Eric Bressler, Rod Martin, and Sigal Balshine, "Production and Appreciation of Humor as Sexually Selected Traits," *Evolution and Human Behavior* 27 (2006): 121–130.

On the role of humor in successful relationships, see William Hampes, "The Relationship Between Humor and Trust," *Humor: International Journal of Humor Research* 12, no. 3 (1999): 253–259; also William Hampes, "Relation Between Intimacy and Humor," *Psychological*

*Reports* 71 (1992): 127–130; also Robert Lauer, Jeanette Lauer, and Sarah Kerr, "The Long-Term Marriage: Perceptions of Stability and Satisfaction," *International Journal of Aging and Human Development* 31, no. 3 (1990): 189–195; also John Rust and Jeffrey Goldstein, "Humor in Marital Adjustment," *Humor: International Journal of Humor Research* 2, no. 3 (1989): 217–223; also Avner Ziv, "Humor's Role in Married Life," *Humor: International Journal of Humor Research* 1, no. 3 (1988): 223–229.

## CHAPTER 8: OH, THE PLACES YOU'LL GO

On the "Malice in Dallas," see Kevin Freiberg and Jackie Freiberg, *Nuts! Southwest Airlines' Crazy Recipe for Business and Personal Success* (Austin, TX: Bard Press, 1996). Actual footage of the match can also be found online.

*Oh, the Places You'll Go*

On humor in the business world, see John Morreall, "Applications of Humor: Health, the Workplace, and Education," in *The Primer of Humor Research,* ed. Victor Raskin (New York: Mouton de Gruyter, 2009).

On humor and organization of public speeches, see John Jones, "The Masking Effects of Humor on Audience Perception and Message Organization," *Humor: International Journal of Humor Research* 18, no. 4 (2005): 405–417.

On humor at West Point, see Robert Priest and Jordan Swain, "Humor and Its Implications for Leadership Effectiveness," *Humor: International Journal of Humor Research* 15, no. 2 (2002): 169–189.

On humor in the classroom, see Robert Kaplan and Gregory Pascoe, "Humorous Lectures and Humorous Examples: Some Effects upon Comprehension and Retention," *Journal of Educational Psychology* 69, no. 1 (1977): 61–65; also Avner Ziv, "Teaching and Learning with Humor: Experiment and Replication," *Journal of Experimental Education* 57, no. 1 (1988): 5–15.

On humor in politics, Congress, and the Supreme Court, see Alan Partington, "Double-Speak at the White House: A Corpus-Assisted Study of Bisociation in Conversational Laughter-Talk," *Humor: International Journal of Humor Research* 24, no. 4 (2011): 371–398;

also Dean Yarwood, "When Congress Makes a Joke: Congressional Humor as Serious and Purposeful Communication," *Humor: International Journal of Humor Research* 14, no. 4 (2001): 359–394; also Pamela Hobbs, "Lawyers' Use of Humor as Persuasion," *Humor: International Journal of Humor Research* 20, no. 2 (2007): 123–156.

On humor and political humility, see Amy Bippus, "Factors Predicting the Perceived Effectiveness of Politicians' Use of Humor During a Debate," *Humor: International Journal of Humor Research* 20, no. 2 (2007): 105–121.

On humor in the workplace, see Barbara Plester and Mark Orams, "Send in the Clowns: The Role of the Joker in Three New Zealand IT Companies," *Humor: International Journal of Humor Research* 21, no. 3 (2008): 253–281; also Owen Lynch, "Cooking with Humor: In-Group Humor as Social Organization," *Humor: International Journal of Humor Research* 23, no. 2 (2010): 127–159; also Reva Brown and Dermott Keegan, "Humor in the Hotel Kitchen," *Humor: International Journal of Humor Research* 12, no. 1 (1999): 47–70; also Leide Porcu, "Fishy Business: Humor in a Sardinian Fish Market," *Humor: International Journal of Humor Research* 18, no. 1 (2005): 69–102; also Janet Bing and Dana Heller, "How Many Lesbians Does It Take to Screw in a Light Bulb?" *Humor: International Journal of Humor Research* 16, no. 2 (2003): 157–182; also Catherine Davies, "Joking as Boundary Negotiation Among Good Old Boys: White Trash as a Social Category at the Bottom of the Southern Working Class in Alabama," *Humor: International Journal of Humor Research* 23, no. 2 (2010): 179–200.

## Greater Implications

On humor and intelligence, see Ann Masten, "Humor and Competence in School-Aged Children," *Child Development* 57 (1986): 461–473.

On humor and insight, see Alice Isen, Kimberly Daubman, and Gary Nowicki, "Positive Affect Facilitates Creative Problem Solving," *Journal of Personality and Social Psychology* 52, no. 6 (1987): 1122–1131; also Heather Belanger, Lee Kirkpatrick, and Peter Derks, "The Effects of Humor on Verbal and Imaginal Problem Solving," *Humor: International Journal of Humor Research* 11, no. 1 (1998): 21–31.

On humor and creativity, see Avner Ziv, "Facilitating Effects of Humor on Creativity," *Journal of Educational Psychology* 68, no. 3 (1976): 318–322.

The finding that watching Robin Williams improves problem-solving ability is from an unpublished paper by Mark Jung-Beeman. For details about the experiment itself, see his December 6, 2010, interview with the *New York Times,* titled "Tracing the Spark of Creative Problem Solving."

*Becoming Funny*

On heritable traits in general, see Matt McGue and Thomas Bouchard, "Genetic and Environmental Influences on Human Behavioral Differences," *Annual Review of Neuroscience* 21 (1998): 1–24. On the heritability of humor specifically, see Beth Manke, "Genetic and Environmental Contributions to Children's Interpersonal Humor," in *Sense of Humor: Explorations of a Personality Characteristic,* ed. Willibald Ruch (New York: Mouton de Gruyter, 1998).

On the comic personality, see Seymour Fisher and Rhoda Fisher, *Pretend the World Is Funny and Forever: A Psychological Analysis of Comedians, Clowns, and Actors* (Hillsdale, NJ: Lawrence Erlbaum Associates, 1981).

On the association between humor comprehension and production, see Aaron Kozbelt and Kana Nishioka, "Humor Comprehension, Humor Production, and Insight: An Exploratory Study," *Humor: International Journal of Humor Research* 23, no. 3 (2010): 375–401.

On humor training, see Ofra Nevo, Haim Aharonson, and Avigdor Klingman, "The Development and Evaluation of a Systematic Program for Improving Sense of Humor," in *The Sense of Humor: Explorations of a Personality Characteristic,* ed. Willibald Ruch (New York: Mouton de Gruyter, 1998).

CONCLUSION

If you really want to read about humor and intoxication, see James Weaver, Jonathan Masland, Shahin Kharazmi, and Dolf Zillmann, "Effect of Alcoholic Intoxication on the Appreciation of Different Types of Humor," *Journal of Personality and Social Psychology* 49, no. 3 (1985): 781–787. On the Russian duo Foma and Yerema, see Alexander Kozintsev, "Foma and Yerema; Max and Moritz; Beavis and Butt-Head: Images of Twin Clowns in Three Cultures," *Humor: International Journal of Humor Research* 15, no. 4 (2002): 419–439.

On the concept of flow, see Mihaly Csikzentmihalyi, *Flow: The Psychology of Optimal Experience* (New York: Harper and Row, 1990).

# INDEX

31901055351284